Closing the Circle
on the
Splitting of the Atom

Earth contaminated with low-level radioactive waste from the Manhattan Project. *Hazelwood Interim Storage Site, Latty Avenue, Hazelwood, Missouri. January 29, 1994.*

Irradiated nuclear fuel in dry storage. *Building 603, Idaho Chemical Processing Plant, Idaho National Engineering Laboratory. March 17, 1994.*

DOE/EM-0266

Closing the Circle on the Splitting of the Atom

The Environmental Legacy
of Nuclear Weapons Production
in the United States
and What the Department of Energy
is Doing About It

The U.S. Department of Energy
Office of Environmental Management

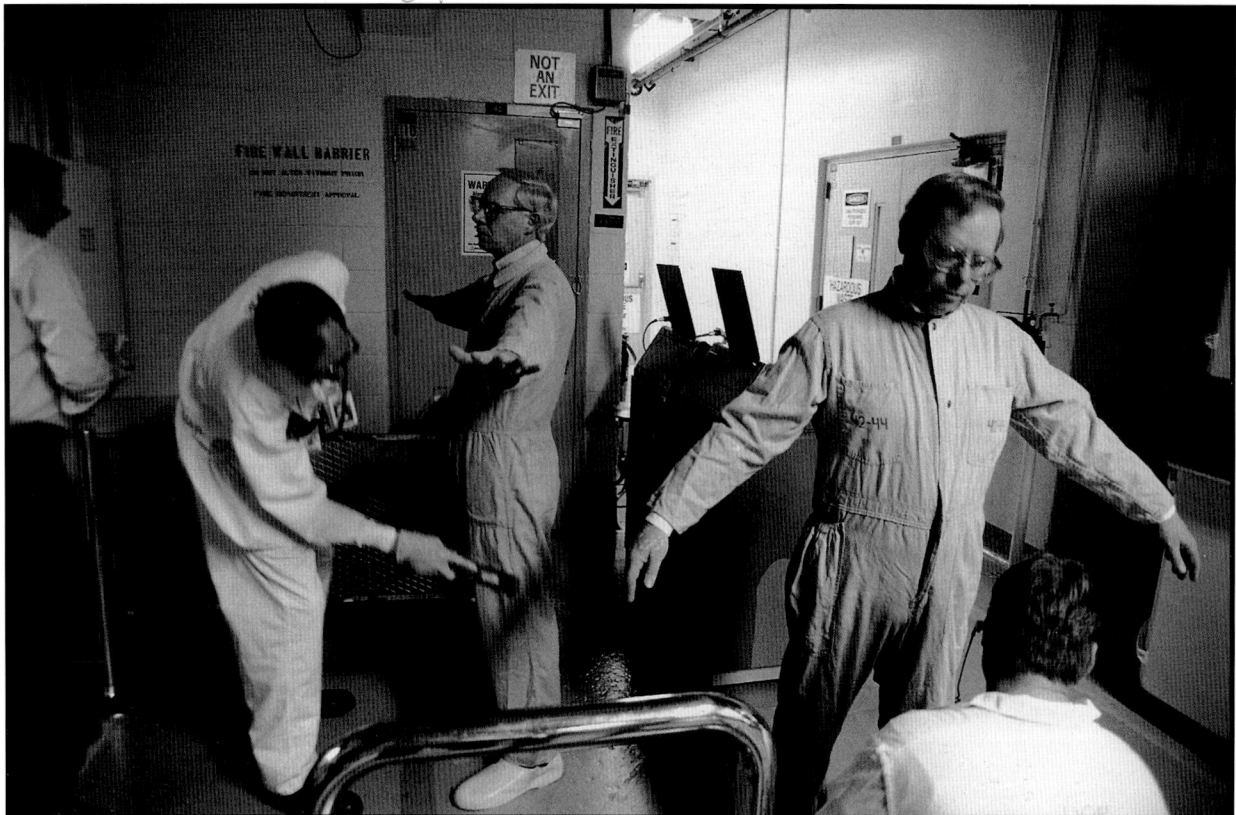

Radiation-safety technicians check workers John Bower and Bill Milligan for possible contamination before they exit a Rocky Flats production building now undergoing cleanup. During the Cold War, the Rocky Flats Plant was the primary facility for processing and machining the plutonium used in nuclear weapons. *Rocky Flats Environmental Technology Site, Colorado. March 19, 1994.*

The photographs on pages 12, 18 (right), 31, and 79 are from AT WORK IN THE FIELDS OF THE BOMB by Robert Del Tredici. Copyright © 1987 by Robert Del Tredici. Reprinted by arrangement with HarperCollins Publishers.

In addition, the photographs on pages 1, 11-14, 16, 37, 39, 58, 59 (number 3), 76, 83, and 100 are copyrighted © by Robert Del Tredici and reprinted with permission by the photographer.

Letter from the Secretary

The United States built the world's first atomic bomb to help win World War II and developed a nuclear arsenal to fight the Cold War. How we unleashed the fundamental power of the universe is one of the great stories of our era. It is a story of extraordinary challenges brilliantly met, a story of genius, teamwork, industry, and courage.

We are now embarked on another great challenge and a new national priority: refocussing the commitment that built the most powerful weapons on Earth towards the widespread environmental and safety problems at thousands of contaminated sites across the land. We have a moral obligation to do no less, and we are committed to producing meaningful results. This is the honorable and challenging task of the Department's Environmental Management program.

Although the war that gave us the atomic bomb ended half a century ago, and the Cold War that followed is now over, the full story of the splitting the atom has yet to be written. **Closing the Circle on the Splitting of the Atom** *reveals one of the story's biggest missing pieces. It describes the environmental legacy of nuclear weapons production in the United States and what the Department of Energy is doing about it.*

This story is being written in laboratories and at waste sites by scientists and engineers grappling with daunting waste and contamination problems. It is being written in state capitols, town halls, and board rooms by government officials, citizens and contractors developing new ways of doing business. And it is being written by tens of thousands of workers dismantling buildings, treating waste, safeguarding plutonium and caring for each other's safety in this dangerous mission.

In 1993 we launched our "Openness Initiative" by coming clean with our past and opening many of our files to the public. We did this to earn public trust and foster informed public participation in Government decisionmaking. This book will help advance this critical obligation by illuminating the challenges and accomplishments of nuclear weapons facilities cleanup and putting a human face on the work being done to close the circle on the splitting of the atom.

Hazel R. O'Leary
Secretary of Energy

The long-range alpha detector is highly sensitive to alpha radiation. This breakthrough in radiation monitoring picks up trace ions formed in air by alpha particles at a distance of several meters, allowing fast screening with quick results. *Los Alamos, New Mexico. February 24, 1994.*

Table of Contents

Letter from the Secretary v
Introduction ix

I. Overview 1
The Manhattan Project 1
The Cold War and the Nuclear Weapons Complex 2
Civilian Control 4
Environmental Legacy of the Cold War 4
Environmental Management 5
Closing the Circle 9
The Challenges Before Us 9

II. Building Nuclear Warheads: The Process 11
Special Nuclear Materials 12
Uranium Mining 13
Uranium Milling 13
Uranium Enrichment 14
Uranium Metallurgy 15
Plutonium Production 16
Extraction of Special Nuclear Materials 18
Plutonium Metallurgy 19
Weapons Design 20
Final Assembly 20
Testing 21

III. Wastes and Other Byproducts of the Cold War 23
Spent Nuclear Fuel 26
Reducing Risks from Spent-Fuel Storage 28
Options for the Long-Term Storage and Disposal of Spent Fuel 29
High-Level Waste from Reprocessing 30

Reducing Risks from High-Level Waste	31
Stabilizing High-Level Waste: Preparing for Disposal	34
Progress in Idaho	35
Converting Waste to Glass in South Carolina, New York, and Washington	36
The Evolution of Health Protection Standards for Nuclear Workers	38
Radiation and Human Health	39
The Plutonium Problem	40
Plutonium Residues and Scraps	41
Plutonium Metal in Storage	42
Informed Debate About Disposition	43
Transuranic Waste	44
Progress in Managing Transuranic Waste	45
Permanent Disposal	45
Low-Level Radioactive Waste	48
Newly Generated Low-Level Waste	49
Managing Low-Level Waste	49
Hazardous Waste	52
Mixed Hazardous and Radioactive Waste	53
Other Materials in Inventory	56
Waste-Handling Complications	58

IV. Contamination and Cleanup 61

Actions in Cleanup	62
Deciding When and How To Take Action	62
Progress in Cleanup	64
Challenges To Be Met	65
Improving Performance	67
Moving Forward	70

V. An International Perspective 75

Worldwide Cooperation	76
Safe Management of Nuclear Materials	76
New Attitudes	77

VI. Transition to New Missions 79

Engineering Challenges	79
Institutional Challenges	80
Need to Know	81
From Secrecy to Openness	81
Whistleblowers	82
Citizen Involvement	82
Contract Reform	82

VII. Looking to the Future 85

Providing for Broad-Based Debate and Participation	85
Strategy Before Action	86
Reconciling Democratic Involvement with Institutional Efficiency	90
The Long-Term Vision for Environmental Management	90
What Might Future Generations Question?	90
Closing the Circle on the Splitting of the Atom	91

Glossary 93
For Further Reading 101

Introduction

In the grand scheme of things we are a little more than halfway through the cycle of splitting the atom for weapons purposes. If we visualize this historic cycle as the full sweep of a clockface, at zero hour we would find the first nuclear chain reaction by Enrico Fermi, followed immediately by the Manhattan Project and the explosion of the first atomic bombs. From two o'clock until five, the United States built and ran a massive industrial complex that produced tens of thousands of nuclear weapons. At half past, the Cold War ended, and the United States shut down most of its nuclear weapons factories.

The second half of this cycle involves dealing with the waste and contamination from nuclear weapons production – a task that had, for the most part, been postponed into the indefinite future. That future is now upon us.

Dealing with the environmental legacy of the Cold War is in many ways as big a challenge for us today as the building of the atomic bomb was for the Manhattan Project pioneers in the 1940s. Our challenges are political and social as well as technical, and we are meeting those challenges. We are reducing risks, treating wastes, developing new technologies, and building democratic institutions for a constructive debate on our future course.

The course of the environmental management program will be decided through broad public debate – both national and local. Where and how will we treat and dispose of the backlog of wastes from nuclear weapons production? How clean is clean? Should we exhume large volumes of contaminated soil in order to allow for unlimited use of the land in the future? Is plutonium a waste or a resource? To foster a sustained and informed public debate on these and other critical questions, we created this book. In it we use photographs as well as facts and figures, because only this combination can begin to convey the scale, the complexity, and the reality of the legacy we face, and the successes we have achieved so far.

Our hope is that this book will promote and inform broad-based citizen involvement so that we can move forward together in this difficult and compelling work.

- *Thomas P. Grumbly*,
Assistant Secretary for Environmental Management, U.S. Department of Energy

I. Overview

Hanford's B Reactor was the first plutonium-production reactor in the world. Plutonium created within this reactor fueled the first atomic explosion in the Alamogordo desert on July 16, 1945, and it formed the core of the bomb that exploded over Nagasaki on August 9, 1945. Built in less than a year, the B Reactor operated from 1944 to 1968. It has been designated a National Historic Mechanical Engineering Landmark. *Hanford Site, Washington. November 16, 1984.*

On a cold morning in December 1989, workers at the Rocky Flats Plant in Colorado loaded the last plutonium "trigger" for a nuclear warhead into a tractor trailer bound southeast to the Pantex Plant near Amarillo, Texas. No one knew then that the nuclear weapon built with this plutonium trigger would be the last one made in the United States for the foreseeable future. Until then, the production of nuclear weapons had run continuously, beginning during World War II with the startup of the first reactor to produce plutonium for the top-secret Manhattan Project. But growing concerns about safety and environmental problems had caused various parts of the weapons-producing complex to be shut down in the 1980s. These shutdowns, at first expected to be temporary, became permanent when the Soviet Union dissolved in 1991. The nuclear arms race of the Cold War came to a halt for the first time since the invention of the atomic bomb. Quietly, a new era had begun.

The Manhattan Project

The quest for nuclear explosives, driven by the fear that Hitler's Germany might invent them first, was an epic, top-secret engineering and industrial venture in the United States during World War II. The term "Manhattan Project" has become a byword for an enormous breakneck effort involving vast resources and the best scientific minds in the world. The workers on the Manhattan Project took on a nearly impossible challenge to address a grave threat to the national security.

Closing the Circle on the Splitting of the Atom

From its beginning with Enrico Fermi's graphite-pile reactor under the bleachers of Stagg Field at the University of Chicago to the fiery explosion of the first atomic bomb near Alamogordo, New Mexico, the Manhattan Project took a little less than 3 years to create a working atomic bomb. During that time, the U.S. Army Corps of Engineers managed the construction of monumental plants to enrich uranium, three production reactors to make plutonium, and two reprocessing plants to extract plutonium from the reactor fuel. In 1939, Nobel Prize-winning physicist Niels Bohr had argued that building an atomic bomb "can never be done unless you turn the United States into one huge factory." Years later, he told his colleague Edward Teller, "I told you it couldn't be done without turning the whole country into a factory. You have done just that."

The Cold War and the Nuclear Weapons Complex

Shortly after World War II, relations between the United States and the Soviet Union began to sour, and the Cold War ensued. Its most enduring legacy was the nuclear arms race. It began during the Manhattan Project, when the Soviet Union began to develop its own atomic bomb.

In the United States, the nuclear arms race resulted in the development of a vast research, production, and testing network that came to be known as "the nuclear weapons complex." Some idea of the scale of this enterprise can be understood from the cost: from the Manhattan Project to the present, the United States spent approximately 300 billion dollars on nuclear weapons research, production, and testing (in 1995 dollars). During half a century of operations, the complex manufactured tens of thousands of nuclear warheads and detonated more than one thousand.

At its peak, this complex consisted of 16 major facilities, including vast reservations of land in the States of Nevada, Idaho, Washington, and South Carolina. In its diversity, it ranged from tracts of isolated desert in Nevada, where weapons were tested, to warehouses in downtown New York that once stored uranium. Its national laboratories in New Mexico and California designed weapons for production in Colorado, Florida, Missouri, Ohio, Tennessee, and Washington. Even now, long after some of the sites used in the nuclear enterprise were turned over to other uses, the Department of Energy–the Federal agency that controls the nuclear weapons complex–owns 2.3 million acres of land and 120 million square feet of buildings.

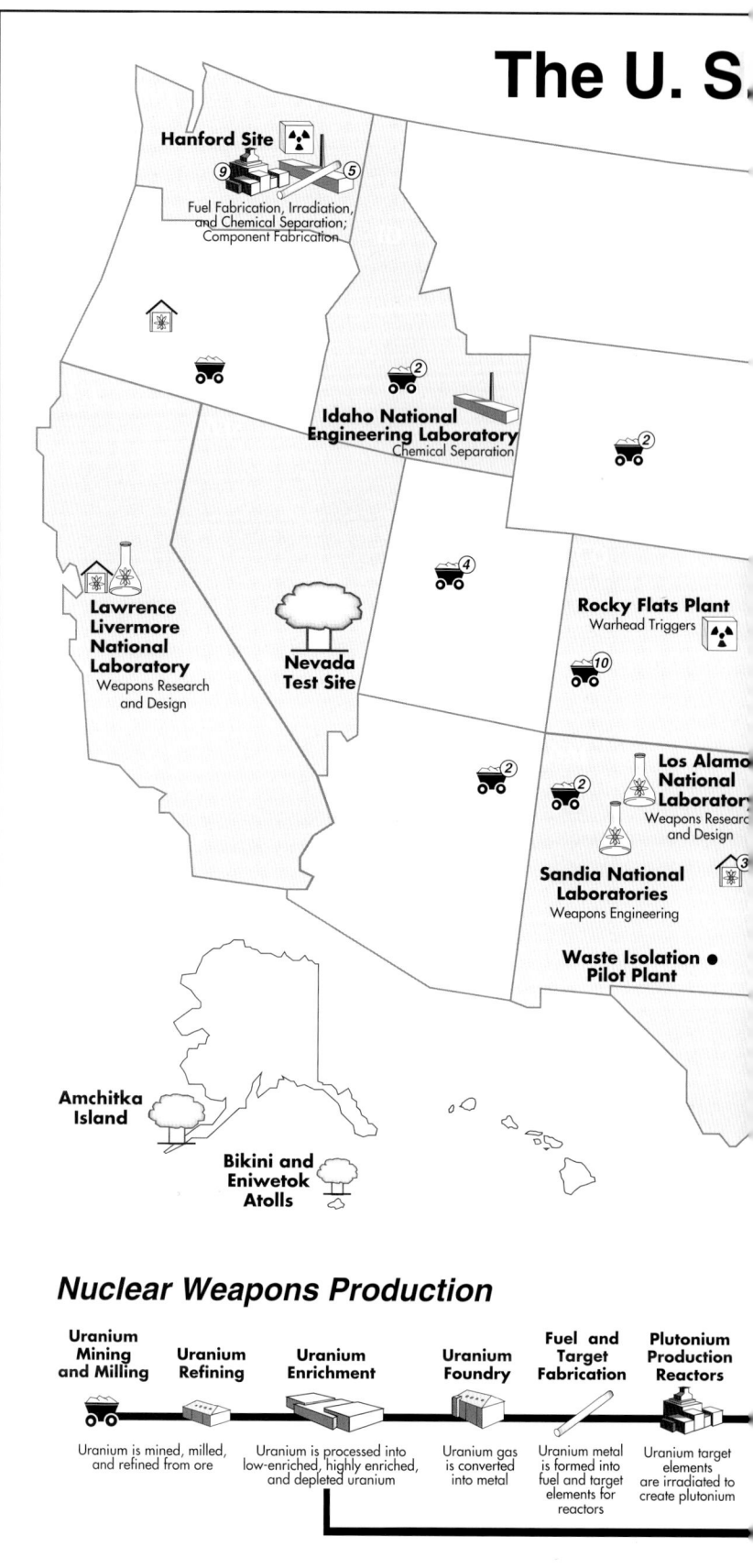

Overview

Nuclear Weapons Complex

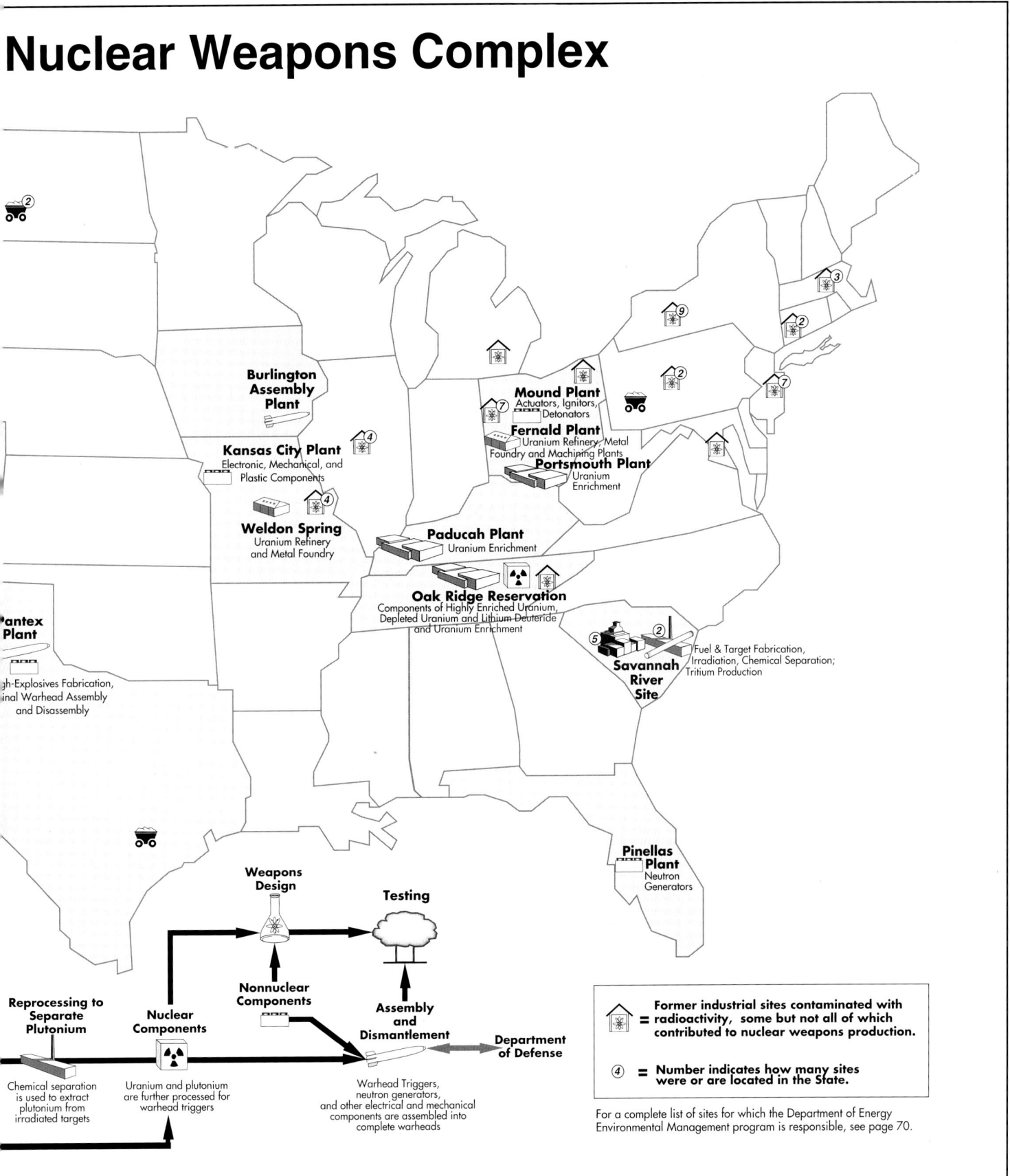

The United States nuclear weapons complex comprised dozens of industrial facilities and laboratories across the country. The weapons production infrastructure originated with the Mahattan Project during World War II and evolved and operated until the late 1980s. It typically employed more than 100,000 contractor personnel at any one time. From the Manhattan Project to the present, the United States has spent approximately $300 billion on nuclear weapons research, production, and testing (in 1995 dollars).

The face of the N Reactor core is made of graphite and measures 39 by 33 by 33 feet. Channels cut horizontally into the graphite held nuclear fuel and uranium "target" slugs. When the slugs were bombarded with neutrons, some of the uranium was transformed into plutonium. During the Cold War the United States operated a total of 14 plutonium-production reactors, creating approximately 100 metric tons of plutonium for its tens of thousands of nuclear warheads. *Hanford Site, Washington. December 16, 1993.*

Civilian Control

Soon after the destructiveness of nuclear weapons was demonstrated by the bombing of Hiroshima and Nagasaki, the U. S. Congress acted to put the immense power and possibilities of atomic energy under civilian control. The Atomic Energy Act of 1946 established the Atomic Energy Commission, to administer and regulate the production and uses of atomic power.

The work of the Commission expanded quickly from building a stockpile of nuclear weapons to investigating peaceful uses of atomic energy (such as research on, and the regulation of, the production of electrical power). It also conducted studies on the health and safety hazards of radioactive materials.

In 1975, the Atomic Energy Commission was replaced by two new Federal agencies: the Nuclear Regulatory Commission, which was charged with regulating the civilian uses of atomic energy (mainly commercial nuclear power plants), and the Energy Research and Development Administration, whose duties included the control of the nuclear weapons complex. In 1977, these duties were transferred to the newly created Department of Energy.

Environmental Legacy of the Cold War

Like most industrial and manufacturing operations, the nuclear weapons complex has generated waste, pollution, and contamination. However, many problems posed by its operations are unlike those associated with any other industry. They include unique radiation hazards, unprecedented volumes of contaminated water and soil, and a vast number of contaminated structures ranging from reactors to chemical plants for extracting nuclear materials to evaporation ponds.

Early in the nuclear age, scientists involved with the weapons complex raised serious questions about its waste-management practices. Shortly after the establishment of the Atomic Energy Commission, its 12-man Safety and Industrial Health Advisory Board reported that the "disposal of contaminated waste in present quantities and by present methods...if continued for decades, presents the gravest of problems."

Overview

The imperatives of the nuclear arms race, however, demanded that weapons production and testing be given priority over waste management and the control of environmental contamination.

Environmental Management

Although the nation continues to maintain an arsenal of nuclear weapons, as well as some production capability, the United States has entered a new era, and the Department of Energy has embarked on new missions. The most ambitious and far-ranging of these missions is dealing with the environmental legacy of the Cold War. In 1989 the Office of Environmental Management was established for that purpose.

Just as the Energy Department's mission of maintaining the nation's nuclear weapons arsenal consists of a number of different tasks, the new mission of Environmental Management involves a variety of interrelated activities. These activities are often generalized simply as "cleanup." In reality, the mission includes four major activities that involve a great deal more than just "cleanup."

> ### Six Goals of Environmental Management
>
> Maintaining surplus facilities, containing radioactive waste, and cleaning up contamination requires a different strategy from weapons production. Assistant Secretary Thomas P. Grumbly has established six goals for the Department of Energy's environmental management program:
>
> - Eliminate and manage urgent risks in our system.
> - Emphasize health and safety for workers and the public.
> - Establish a system that is managerially and financially in control.
> - Demonstrate tangible results.
> - Focus technology development on identifying and overcoming obstacles to progress.
> - Establish a stronger partnership between the Department of Energy and its stakeholders.

Barrels of transuranic waste sit on a concrete pad in temporary storage. This waste is contaminated with traces of plutonium, which is dangerous if inhaled and will remain a hazard for hundreds of thousands of years. More than 300,000 barrels of such waste from nuclear weapons production are buried or stored around the country. Cleanup efforts throughout the weapons complex will add to the volume of this waste. *Transuranic Waste Storage Pads, E Area Burial Grounds, Savannah River Site, South Carolina. January 7, 1994.*

Closing the Circle on the Splitting of the Atom

Empty drums used for storing waste await treatment and disposal at Oak Ridge. These drums corroded prematurely when a 1987 waste-stabilization project failed to follow guidelines for combining waste sludge with cement. *K-1417 Drum Storage Yards, Pond Waste Management Project, Oak Ridge, Tennessee. January 10, 1994.*

The first major activity is managing *urgent and high-risk nuclear materials and facilites.* For example, the reprocessing plants are no longer needed for the extraction of weapons-grade plutonium, and the nuclear materials inside are not intended to be used for nuclear weapons. The task of stabilizing these facilities and the extraordinarily sensitive material inside them to prevent leaks, explosions, theft, terrorist attack, or avoidable radiation exposures is part of the mission of Environmental Management. Maintaining these facilities has become more difficult because many of them are more than 40 years old. Many have reached or exceeded the lifetime they were designed for and have begun to deteriorate; they must be stabilized merely to protect the safety of cleanup workers. This stable condition must be achieved and the facilities and material must be kept in a safe condition before any decontamination and decommissioning can be undertaken.

Environmental Management also supports international nuclear nonproliferation policies. Specifically, spent-fuel elements removed from reactors were recently returned from other countries to the United States because they contained weapons-grade uranium of U.S. origin. The United States thereby reduced the international trade of weapons-usable highly enriched uranium.

The second major activity is *managing a large amount and variety of wastes*. The primary source of these wastes is the nuclear weapons activities of the Cold War. In addition, the Department also manages some waste from nuclear reactor research and basic science projects, as well as some waste generated by the commercial nuclear power industry under certain circumstances, such as the debris from the accident at the Three Mile Island reactor. Most of the waste generated by the Energy Department is radioactive, and therefore cannot be eliminated – it can only be contained while its radioactivity diminishes. A large volume of waste has already been disposed of at Department of Energy facilities. However, the wastes that remain in storage pending permanent disposal contain most of the radioactivity. These wastes, which will typically remain hazardous for thousands of years, are intended for deep geologic disposal. Part of the task of the Office of Environmental Management is to conduct the scientific investigations required to determine the suitability of a deep salt mine already excavated in New Mexico for plutonium-contaminated waste.

Overview

In addition, waste management includes designing, building, and operating a variety of treatment facilities to prepare waste for disposal. Providing safe storage for the enormous quantities of waste is itself a monumental challenge. At the Hanford Site, for example, the Department maintains a constant vigil over huge underground tanks of highly radioactive waste, and it has recently installed a pump in one tank that was at risk of exploding.

The third major activity, *environmental restoration*, is the activity that is usually visualized when the program is described simply as "cleanup." This part of the program encompasses a wide range of activities, including stabilizing contaminated soil; pumping, treating, and containing ground water; decontaminating, decommissioning, and demolishing process buildings, nuclear reactors, and chemical separation plants; and exhuming sludge and buried drums of waste. The challenges are both technical and institutional. In many cases, no safe or effective technology is yet available to address – or even fully understand – the contamination problem. Choosing the right course of action requires the involvement of

Environmental Management includes four major activities that involve much more than just "cleanup."

"Pit Nine" is a radioactive-waste burial ground. From 1967 to 1969, approximately 150,000 cubic feet of plutonium-contaminated and low-level radioactive waste was buried here. Recordkeeping that does not meet today's standards, and failed waste containment have made Pit Nine a daunting remediation challenge for engineers, who must now sample these wastes, exhume them, and treat them thermally. *Radioactive Waste Management Complex, Idaho National Engineering Laboratory. March 16, 1994.*

REPORT OF THE SAFETY AND
INDUSTRIAL HEALTH
ADVISORY BOARD

April 2, 1948

The Atomic Energy Commission isolated its projects, built plants which are a marvel of engineering and guarded them with extraordinary efficiency. Their sins of emission—liquid, solid, or gaseous—were diluted and isolated to what was estimated as perfectly safe, but AEC is now entering a phase in which their operations in this regard will soon be public property and they will be accountable to public health—a very severe critic...

In the haste to produce atomic bombs during the war certain risks may have been taken in research, production, testing, transportation and waste disposal with the understanding that subsequently more effective control measures would ameliorate these risks and lessen the hazardous conditions formerly created...

The ultimate disposal of contaminated waste—sub-surface, surface and airborne—needs much more thorough study. Even the simplest of such data—recorded periodic measurements of stream pollution below the plants—are almost wholly lacking. Even with such records, present knowledge of radiation and chemically toxic effects on animal and vegetable life is so limited that water supply inlets below plant disposal outlets cannot be unqualifiedly recommended. The disposal of contaminated waste in present quantities and by present methods (in tanks or burial grounds or at sea), if continued for decades, presents the gravest of problems.

– from pages 9, 64, 67

environmental regulatory agencies, State and local governments, and the general public. Where possible, contaminated buildings and equipment are restored to prepare them for other uses. The main objectives are to avoid additional problems, minimize hazards to workers and the public, and minimize the cost and risks passed on to future generations.

The fourth major activity, *technology development*, is perhaps the most vital to the long-term success of the environmental management mission. The Energy Department is conducting a variety of applied research to develop more effective and less expensive remedies to the environmental and safety problems of the nuclear weapons complex. Some of this research has already yielded signficant results. A good example is a technique, known as Minimum Additive Waste Stabilization, that was demonstrated at the Fernald site in Ohio to convert low-level radioactive waste into flattened glass pebbles, which are easy to handle and will remain stable after disposal. The success of this research is demonstrated not only by improvements in environmental protection but also by the commercialization of these technologies.

Solving the problems of the Cold War's environmental legacy will take many decades, enormous financial resources, and continued guarding and monitoring of sites.

A temporary tension-support structure is being constructed at Fernald. These lightweight structures are increasingly used throughout the former nuclear weapons complex. They keep drums of various types of waste out of the elements, extending their storage life at relatively low cost. *Plant 1 pad, Fernald Environmental Management Project, Fernald, Ohio. December 28, 1993.*

Closing the Circle

The Cold War is over, but its legacy remains. Solving the waste-management and contamination problems of this legacy will take many decades and hundreds of billions of dollars. Even then the task will not be fully completed. Many sites and facilities will need continued guarding and monitoring.

In speaking about the evolution of life on earth, scientist Barry Commoner said:

The first photosynthetic organisms transformed the ...linear course of life into the...first great ecological cycle. By closing the circle, they achieved what no living organism alone can accomplish–survival. Once the links between the separate parts of the problem are perceived, it becomes possible to see new means of solving the whole.

The task of Environmental Management is to begin to close the circle on the splitting of the atom for weapons production through sustained efforts to understand the whole problem as well as its parts.

The Challenges Before Us

The nation faces daunting institutional and technical challenges in dealing with the environmental legacy of the Cold War. We have large amounts of radioactive materials that will be hazardous for thousands of years; we lack effective technologies and solutions for resolving many of these environmental and safety problems; we do not fully understand the potential health effects of prolonged exposure to materials that are both radioactive and chemically toxic; and we must clear major institutional hurdles in the transition from nuclear weapons production to environmental cleanup.

These problems cannot be solved by science alone. In the midst of the complexities and uncertainties, one thing is clear: the challenges before us will require a similar–if not greater–level of commitment, intelligence, and ingenuity than was required by the Manhattan Project.

Closing the Circle on the Splitting of the Atom

II. Building Nuclear Warheads: The Process

Haystack Mountain, near Grants, New Mexico, is the richest uranium-mining district in the United States. After uranium deposits were discovered here in 1950, many large underground and open-pit mines were opened in the area, and some operations continued until 1990. *Grants, New Mexico. August 19, 1982.*

The production of nuclear weapons requires special technologies that were invented for the Manhattan Project. It also requires special materials: highly enriched uranium and plutonium. Both are made, by different processes, from naturally occurring uranium ore. Mining uranium ore is thus the first link in a chain of complex processes that eventually produce a nuclear weapon.

Although plutonium and uranium are both essential parts of modern nuclear weapons, it is possible to make nuclear weapons by using one or the other material alone. In fact, the first generation of atomic weapons did so. Early nuclear weapons were of two types: (1) gun-type bombs using two masses of highly enriched uranium, forced together very quickly to assemble a "critical mass" that would sustain a nuclear chain reaction and subsequent explosion; and (2) implosion bombs using high explosives to squeeze together a sphere of plutonium very quickly and symmetrically into a critical mass to attain a nuclear explosion. The "Little Boy" bomb dropped on Hiroshima was a uranium gun-type weapon, while the bomb dropped on Nagasaki was a plutonium implosion bomb. As designs for nuclear weapons advanced, a new generation of bombs – thermonuclear weapons – evolved. Most modern nuclear weapons use both plutonium and uranium.

Tritium is another essential material in most nuclear weapons. It is a radioactive gas that is produced by bombarding lithium with neutrons in a reactor. It is used to boost the explosive power of many modern weapons.

Closing the Circle on the Splitting of the Atom

A model of a uranium atom is displayed at the American Museum of Science and Energy in Oak Ridge. Uranium is the basic element from which nuclear explosives are made. *Oak Ridge, Tennessee. June 11, 1982.*

Special Nuclear Materials

In nature, more than 99 percent of the atoms in uranium have an atomic weight of 238. From this, the remaining 1 percent, a particular atomic form, or isotope, with a weight of 235 must be physically separated in sufficient quantities to sustain a nuclear chain reaction–either for generating electrical power or, at much higher concentrations, for explosives.

Separating sufficient quantities of uranium 235 requires enormous amounts of energy and the meticulous operation of large, complex facilities. During the Manhattan Project, two separation methods were pursued simultaneously: electromagnetic separation in the "Calutron" (California University Cyclotron) and gaseous diffusion. Facilities for both methods were built at the Oak Ridge Reservation in Tennessee.

Since then, however, gaseous diffusion has generally been used in the United States to enrich uranium. The process involves a series of vast structures designed to drive gaseous uranium at controlled temperatures and pressures through miles of filters that gradually collect uranium 235 atoms in increasing concentrations– a process called "uranium enrichment." Two additional diffusion plants were built in Ohio and Kentucky in the 1950s.

Highly enriched uranium (more than 20 percent uranium 235, and typically more than 90 percent) is used in nuclear weapons. Low-enriched uranium, consisting of less than 20 percent uranium 235, is nearly impossible to make bombs with, but is used as fuel for nuclear reactors. The uranium 238 that is removed in the enrichment process is called "depleted uranium." It is used in some nuclear weapon parts as radiation shielding, in tank armor, and in armor-piercing bullets. It is also used to make plutonium.

Scientists knew they could avoid the trouble of enriching uranium if they could produce another nuclear material that could be chemically separated from impurities for use in bombs. That material was plutonium 239–an element that is created in nuclear reactors. In the nuclear fuel for a production reactor, uranium 235 is split into a host of radioactive byproducts; in the process, it releases neutrons. The neutrons bombard the uranium 238 in the fuel and transform it into the heavier element, plutonium 239.

Plutonium, like uranium, is a mix of several isotopes. Material rich in the isotope plutonium 239 is referred to as "weapons-grade plutonium."

After plutonium 239 has been created in the reactor, workers must separate it from the uranium and the radioactive byproducts (fission products) in a reprocessing plant. This plant dissolves irradiated uranium in acid and then extracts the uranium and plutonium, leaving behind a highly radioactive liquid referred to as high-level waste.

Because radiation levels inside a reprocessing plant are very high, the plant must be heavily shielded and operated by remote control to protect workers and the environment.

The Trinity nuclear test took place on this spot in the Alamogordo desert of New Mexico. In the background, a stone monument identifies the epicenter of the world's first nuclear explosion. In the foreground is a duplicate outer casing of the Nagasaki bomb. Both the Trinity and the Nagasaki bombs used plutonium cores. *White Sands Missile Range, Alamogordo desert, New Mexico. July 16, 1985.*

Uranium Mining

Most of the uranium for the Manhattan Project came from rich deposits in Africa and Canada, but more than 400 mines eventually opened in the United States, primarily in Arizona, Colorado, New Mexico, Utah, and Wyoming. After World War II, uranium mining expanded dramatically, from 38,000 tons of ore in 1948 to 5.2 million tons in 1958 – nearly all of it for nuclear weapons production. The United States mined about 60 million tons of ore to produce this uranium. Many tons of natural uranium were needed to produce the several kilograms of enriched uranium used in the Hiroshima bomb. For each kilogram of plutonium made for the U.S. arsenal, miners took roughly 1,000 tons of uranium ore from the ground.

Uranium Milling

A ton of uranium ore yields only a few pounds of uranium metal. The result is a dry purified concentrate called "yellowcake." The milling produces large volumes of a sandlike byproduct called "mill tailings." These tailings contain both toxic heavy metals and radioactive radium and thorium. Uranium – mill tailings account for a small fraction of the radioactivity in the byproducts of weapons production, but they constitute 96 percent of the total volume of radioactive byproducts for which Environmental Management is responsible. Because uranium mills typically piled tailings without covers or containment, some material was spread by wind and water. The primary hazard of these tailings is the emission of the radioactive gas radon. The Congress passed a law in 1978 to ensure that these tailings would be adequately stabilized.

Most of the uranium for the Manhattan Project came from from Africa and Canada. Later more than 400 mines opened in the United States.

Converter vessels in a gaseous-diffusion plant contain porous barriers that enrich uranium in gaseous form by separating out the atoms of uranium 235 from more-abundant uranium 238. Each of these vessels is a stage in the enrichment process, and there are a total of 5,122 stages at this plant. The more stages uranium hexafluoride gas passes through, the higher its enrichment becomes. *Unit 7, Cell 2, K-33 Demonstration Cell, K-25 Site, Oak Ridge, Tennessee. June 21, 1993.*

Uranium mills shipped the yellowcake to plants that refined the concentrate into forms suitable for several different roles in weapons production. Metallic uranium was used as fuel in the plutonium-production reactors at Hanford and at the Savannah River Plant. The Fernald Plant in Ohio was the principal site where many thousands of tons of uranium were refined, and sent to the enrichment plants at Oak Ridge, Tennessee; Paducah, Kentucky; and Portsmouth, Ohio, to be used in processes that separated and concentrated the uranium 235.

Uranium Enrichment

To make highly enriched uranium, the enrichment plants used an elaborate process to separate most of the rare uranium 235 from the more abundant uranium 238 isotope. The U.S. Government used most of the highly enriched uranium produced between 1943 and 1964 to make nuclear weapons.

The government made additional highly enriched uranium, with an enrichment of 20 to 90 percent, until 1992. The highly enriched uranium not used in weapons has been used primarily as a fuel for plutonium-production reactors or naval propulsion reactors. Smaller quantities have been used in research reactors. The government plants made a total of 994 tons of highly enriched uranium.

The vast majority of the material fed into the enrichment plants came out as depleted uranium, also called enrichment "tails." Many thousands of tons of depleted uranium are still stored in cylinders in Ohio, Tennessee, and Kentucky. Moreover, operations at enrichment plants over the years caused extensive environmental contamination with solvents, polychlorinated biphenyls (PCBs), heavy metals, and other toxic substances.

The Process

Uranium Metallurgy

Uranium is converted into metal before it is used in nuclear weapons production. Workers at the Fernald uranium foundry in Ohio converted hundreds of tons of uranium hexafluoride gas (the "tails" from the enrichment process) into uranium "green-salt" crystals. These crystals were blended with magnesium granules and cooked in a furnace. The mixture ignites, converting the green-salt crystals into uranium metal. Some of this metal was made into reactor fuel or target elements for plutonium production reactors at Hanford and Savannah River. The Rocky Flats Plant in Colorado and the Y-12 Plant in Oak Ridge, Tennessee, formed depleted and enriched uranium metal into components for nuclear weapons. Releases of uranium dust and leaking landfills of chemicals were the primary environmental impacts of these operations.

Many thousands of tons of depleted uranium are stored in cylinders in Ohio, Tennessee, and Kentucky.

Uranium hexafluoride cylinders are stored near the K-25 Gaseous Diffusion Plant. Each cylinder contains 10 to 14 tons of depleted uranium hexafluoride. The K-25 plant stores about 5,000 of these cylinders, some of them 40 years old. Here, a worker is using ultrasound to evaluate the effects of external corrosion on a steel cylinder. Cylinders closest to the ground can experience accelerated corrosion. *K-1066K Cylinder Storage Yard, K-25 Site, Oak Ridge, Tennessee. January 9, 1994.*

Closing the Circle on the Splitting of the Atom

Final inspection of uranium "billets." These billets of depleted uranium metal were produced at the Fernald Plant to be used as the cores of Mark 31 targets. Jack Schick, a metals worker, conducts a final inspection of a new batch before shipping them to the Savannah River Site, where they would be clad in aluminum, bombarded with neutrons, and partly transformed into plutonium. *Plant 6, Fernald Feed Materials Production Center, Fernald, Ohio. December 17, 1985.*

To produce plutonium, workers at the Hanford and Savannah River Sites processed hundreds of thousands of tons of uranium.

Plutonium Production

Between 1944 and 1988, the United States built and operated 14 plutonium-production reactors at the Hanford and Savannah River Sites, producing a total of about 100 metric tons of plutonium.

After the uranium from Fernald was coated with aluminum or zirconium metal, it was assembled into reactor fuel and targets. The Hanford Site's nine reactors all consist of large cubes of graphite blocks with horizontal channels cut in them for the uranium fuel and cooling water. The fuel slugs were inserted into the front face of the reactor where they underwent neutron bombardment. Then they were gradually pushed through the channels until they fell out the other side. The Savannah River Site's five reactors are different. They each consist of a large tank of "heavy water" in which highly enriched fuel and separate depleted-uranium targets were submerged. Because only a small fraction of the uranium in fuel and targets was converted to plutonium during each cycle through a reactor, workers at Hanford and Savannah River processed hundreds of thousands of tons of uranium. The production reactors at Savannah River also made tritium.

Jack Weaver: A Worker at Rocky Flats

Jack Weaver is one of the few old-timers working at the Rocky Flats Environmental Technology Site, the former nuclear weapons plant in Colorado. Only about two dozen of the 7,000 employees now at Rocky Flats have been there longer than Weaver. The accelerating exodus of experienced workers from Rocky Flats has led Weaver to take on the role of teacher and site historian.

"I'm concerned about the number of people we've lost over the past few years and the knowledge they've taken with them," says Weaver. As the manager of operations of one of the oldest plutonium processing buildings at the site, Weaver has seen it all. He worked there when Rocky Flats received plutonium from other sites to purify and fashion it into triggers for nuclear war. He has seen the consequences of mishandling plutonium. He saw plutonium fires. He knows safety rules are not merely paper exercises.

Weaver began his career at Rocky Flats in 1961 at the age of 20. "They told me it was a good place to work, that the pay and benefits were good," he says. At the time he started, there were 1,430 workers at the site. Within a year, increased production requirements raised that number to about 2,500. Weaver became a chemical operator and then worked his way up to operations manager.

"We went from a couple shifts, five days a week, to around the clock, seven days a week," Weaver says. "The Cold War was on and we all felt we had an important job to do for the country. We were all proud of what we were doing."

As proud as they were, however, the workers couldn't do much boasting. "Nobody talked about what we were doing in those days," Weaver says. "We didn't even talk about it amongst ourselves, let alone with our families and friends offsite. On the floor even, we referred to "Y" or "Z" or "U" and not to beryllium or plutonium or enriched uranium."

"So a lot has changed from those days, now that the Cold War and production mission are over and with the new openness policies," Weaver says. "The biggest change at the site came with the curtailed operations in 1989 after we were shut down."

With the Department's new openness, Weaver is able to dedicate himself to educating people about the site. "I now see myself as a teacher and guide, letting people know what we did here during the days of the Cold War, so that maybe they will have an understanding of what went on," Weaver says. "I've seen a lot in 33 years, and it should be of good use to someone."

The head of the K Reactor is seen here through a 4-foot-thick window of lead-glass shielding. Uranium-metal targets were placed inside the reactor and bombarded with neutrons to convert the uranium to plutonium. This reactor also bombarded lithium targets to make tritium, a gas used to boost the explosive power of nuclear weapons. *Savannah River Site, South Carolina. January 7, 1994.*

Closing the Circle on the Splitting of the Atom

This tritium facility extracted tritium gas from lithium reactor targets. *Savannah River Site, South Carolina. January 7, 1994.*

The B Plant canyon was the world's second large-scale reprocessing facility. It dissolved irradiated fuel rods in acid to recover plutonium. *Hanford Site, Washington. November 15, 1984.*

Extraction of Special Nuclear Materials

The irradiated fuel and targets discharged from production reactors contained hundreds of different radioactive isotopes, collectively called "fission products." These had to be separated from the uranium and plutonium. Scientists developed chemical processes to accomplish this separation. Because exposure to even small amounts of these fission products would be lethal in a short time, workers could handle them only by remote control behind lead-glass shielding and thick concrete walls. In the United States, eight of these chemical separation plants, called "canyons," were operated for recovering plutonium and uranium until the late 1980s. For example, the PUREX facility at the Hanford Site in Washington operated from 1956 to 1972 and resumed operation from 1983 through December 1988. Plants were also operated in South Carolina and Idaho.

Reprocessing plants have generated 105 million gallons of highly radioactive and hazardous chemical waste–enough to fill a 1,000-foot-long supertanker. High-level reprocessing wastes contain almost 99 percent of the total radioactivity left from nuclear weapons production. They also contain long-lived radioactive elements that could pose environmental risks for tens of thousands of years. Reprocessing also generated billions of gallons of wastewater. Although this wastewater contained only about 1 percent of the radioactivity and trace amounts of chemicals, it caused widespread contamination because it was discharged directly to the ground during the Cold War.

The Process

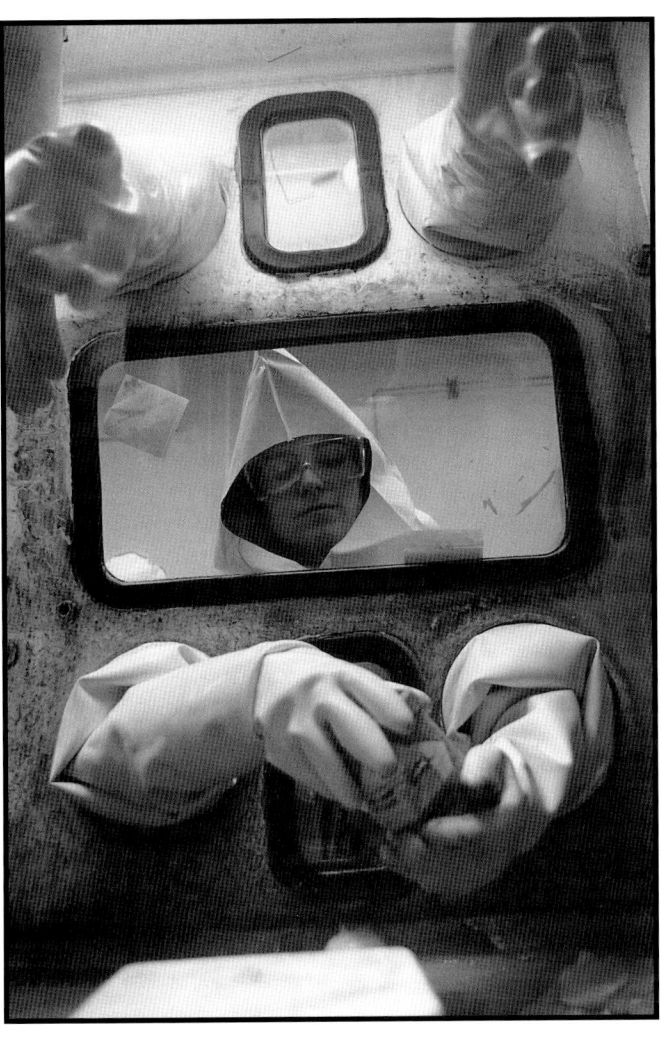

A glovebox for handling plutonium is a sealed environment kept under negative pressure and, when necessary, filled with inert gas to keep the plutonium inside it from igniting in air. Safety procedures require this plutonium worker to wear anticontamination clothing and to handle plutonium through rubber gloves attached to the wall of the box. *Plutonium Finishing Plant, Hanford Site, Washington. December 17, 1993.*

Plutonium Metallurgy

Most plutonium from the reprocessing plants went to the Rocky Flats Plant in Colorado to be machined into warhead components. It was usually in the form of a metal, but liquid and powdered forms of plutonium were also produced. The weapons laboratories used some plutonium to make and test prototype designs for weapons.

Plutonium can be extremely dangerous, even in tiny quantities, if it is inhaled. Because of these hazards, plutonium metallurgy required workers to use gloveboxes equipped with safety and ventilation systems.

Calibration spheres like this one surveyed by Y-12 machinist Danny Bush were used to set instruments that check the manufacturing specifications for weapons parts. Crucial components of nuclear warheads must be accurately shaped to within a few millionths of an inch. *Y-12 Metrology Laboratory, User Facility Skills Demonstration Center, Y-12 Plant, Oak Ridge, Tennessee. January 12, 1994.*

Closing the Circle on the Splitting of the Atom

An example of a completed nuclear weapon and its component parts. At top, an intact B-61 nuclear bomb. At bottom, the assemblies and subassemblies that comprise this weapon. Dozens of facilities across the country engage in different processes and contribute specific parts to the production of nuclear weapons.

Weapons Design

Research, development, and testing have been a critical part of the nuclear weapons enterprise. Two national laboratories – at Livermore, California, and Los Alamos, New Mexico – devoted their expertise to this work during the Cold War. A third laboratory – Sandia National Laboratories, based in Albuquerque, New Mexico – worked on the electronic mechanisms for nuclear warheads as well as designs for coupling the warheads to bombs and missiles. Many different types of nuclear bombs and warheads have been manufactured in the United States, and some additional designs were partially developed.

Final Assembly

Factories in several States (Florida, Missouri, Ohio) contributed components for the final assembly of nuclear weapons. Final assembly occurred primarily near Amarillo, Texas, at the Pantex Plant. The assembly process did not create much radioactive waste.

With the end of the Cold War, the Department of Energy has reversed the activities at Pantex. The plant now disassembles warheads that have been retired from the nation's arsenal, and it is now storing most of their plutonium components. Uranium components are shipped to Oak Ridge, Tennessee. Tritium canisters are shipped to the Savannah River Site in South Carolina. Interim storage and ultimate disposition of surplus nuclear weapons materials pose a number of challenges, such as worker and public safety and security against potential theft.

The Process

"Gravel gerties" are concrete structures whose roofs consist of cable mesh supporting large amounts of gravel. Beneath them are bays, where workers assemble and disassemble nuclear warheads. Should a warhead's conventional explosives accidentally detonate, the roofs of these structures are engineered to give way, releasing the gravel and trapping the plutonium particles. Up to 2,000 warheads per year are now being dismantled at this site. *Pantex Plant, Amarillo, Texas. November 18, 1993.*

Testing

During the past 50 years, the United States exploded more than 1,000 individual nuclear devices in atmospheric, underwater, and underground tests. Most of the nuclear weapons tests were conducted in Nevada, but tests were also done in the Pacific Ocean, Alaska, the south Atlantic, and New Mexico. Nuclear explosion tests were also conducted in Colorado, New Mexico, Mississippi, and Alaska for non-weapons purposes. These tests were done to explore the potential use of nuclear explosions to extract natural gas or to dig harbors. Radioactive contamination from testing remains at most of the test sites.

The United States stopped atmospheric testing in 1963 and has not conducted any nuclear explosion tests since September, 1992.

The Pantex Plant, near Amarillo, Texas, now disassembles nuclear warheads and is storing their plutonium components.

Closing the Circle on the Splitting of the Atom

III. Wastes and Other Byproducts of the Cold War

A 55-gallon drum ready for storage in the basement level of the K-25 Gaseous Diffusion Plant. K-25 has prepared 45 basement vaults for the storage of low-level and mixed hazardous wastes. These vaults will be able to hold some 63,000 drums. A coat of epoxy sealant covers the renovated vault floor, adding one more level of containment for wastes stored here. *K-25 Gaseous Diffusion Plant, Oak Ridge, Tennessee. January 10, 1994.*

Every step in the production of materials and parts for nuclear warheads generated waste and other byproducts. Every gram of plutonium, each reactor fuel element, every container of enriched uranium, and each canister of depleted uranium has radioactive waste associated with it. The graphite bricks used by Enrico Fermi for his primitive reactor at the University of Chicago were buried as radioactive waste at the Palos Forest Preserve in Cooke County, Illinois. The acid used to extract the plutonium for the first nuclear test explosion in the Alamogordo desert of New Mexico is now high-level waste stored at the Hanford Site in the State of Washington.

The wastes are classified into several categories, depending on the hazards they pose, the length of time they remain radioactive, or their source. They require safe storage and disposal, and they often need special treatment before either storage or disposal.

The nuclear weapons industry typically used waste-disposal methods that were considered acceptable at the time–especially between 1943 and 1970. By today's standards, however, these methods would be considered primitive. One result of these practices is significant contamination of soil and ground water (see Chapter IV). For example, some types of liquid waste were held in ponds for evaporation because engineers did not expect radioactive material to seep into the soil and ground water as rapidly as it did.

Every step in the production of materials and parts for nuclear warheads generated waste and other byproducts.

Nuclear weapons wastes are as varied as the processes that produced them: intensely radioactive acids from reprocessing; slightly radioactive shoe covers from walking across factory floors; chemical solvents from performing purity tests. Each of these wastes differs in physical characteristics, chemical form (salt-cake, acidic liquid), and radioactivity (short-lived tritium, long-lived plutonium). Each requires different handling and the volume of waste continues to grow. Every time workers suit up and walk into a contaminated building for an inspection, they create more waste (gloves, shoe covers, disposable coveralls). Each time a ventilation system is cleaned out, waste is the result. Sampling excavated radioactive solids creates waste. The process of stabilizing and cleaning up old facilities generates huge volumes of additional waste.

The radioactivity level of all this waste is slowly decreasing. With the shutdown of the last production reactor in 1988, the total amount of radioactivity in the system stopped growing and is now decreasing at the decay rate of the various isotopes. Some isotopes decay quickly, with half-lives of only a few minutes, others have half-lives of many thousands of years.

This chapter follows the path of major process materials through the complex. It starts with a discussion of spent fuel, then considers highly radioactive waste from chemical separation. Next comes a discussion of plutonium, then of transuranic waste. The chapter continues with sections on low-level waste, hazardous waste, mixed radioactive – and – hazardous waste, and finally materials left in the inventory that were once used in production but no longer have a clearly identified use. Uranium-mill tailings are considered to be contamination rather than waste and are discussed in the chapter on contamination.

Categories of Radioactive Wastes and Byproducts

The Department is responsible for managing large inventories of nuclear waste and byproducts in accordance with national and international principles. These principles require protection of the environment and health for present and future generations, compliance with independent regulatory agencies, and a practicable minimum of waste generation. The primary waste and byproduct categories are defined as follows:

Spent fuel: fuel elements and irradiated targets (designated "reactor-irradiated nuclear material" and often called simply "spent fuel") from reactors. The Department's spent fuel is not categorized as waste, but it is highly radioactive and must be stored in special facilities that shield and cool the material.

High-level waste: material generated by the reprocessing of spent fuel and irradiated targets. Most of the Department's high-level waste came from the production of plutonium. A smaller fraction is related to the recovery of enriched uranium from naval reactor fuel. This waste typically contains highly radioactive, short-lived fission products as well as long-lived isotopes, hazardous chemicals, and toxic heavy metals. It must be isolated from the environment for thousands of years. Liquid high-level waste is typically stored in large tanks, while waste in powdered form is stored in bins.

Transuranic waste: waste generated during nuclear weapons production, fuel reprocessing, and other activities involving long-lived transuranic elements. It contains plutonium, americium and other elements with atomic numbers higher than that of uranium. Some of these isotopes have half-lives of tens of thousands of years, thus requiring very long-term isolation. Since 1970 transuranic waste has been stored temporarily in drums at sites throughout the complex.

Low-level waste: any radioactive waste that does not fall into one of the other categories. It is produced by every process involving radioactive materials. Low-level waste spans a wide range of characteristics, but most of it contains small amounts of radioactivity in large volumes of material. Some wastes in this category (e.g., irradiated metal parts from reactors) can have more radioactivity per unit volume than the average high-level waste from nuclear weapons production. Most low-level waste has been buried near the earth's surface. A limited inventory remains stored in boxes and drums.

Mixed waste: waste that contains both radioactive and chemically hazardous materials. All high-level and transuranic waste are managed as a mixed waste. Some low-level waste is mixed-waste.

Uranium-mill tailings: large volumes of material left from uranium mining and milling. While this material is not categorized as waste, tailings are of concern both because they emit radon and because they are usually contaminated with toxic heavy metals, including lead, vanadium, and molybdenum.

Byproducts of the Cold War

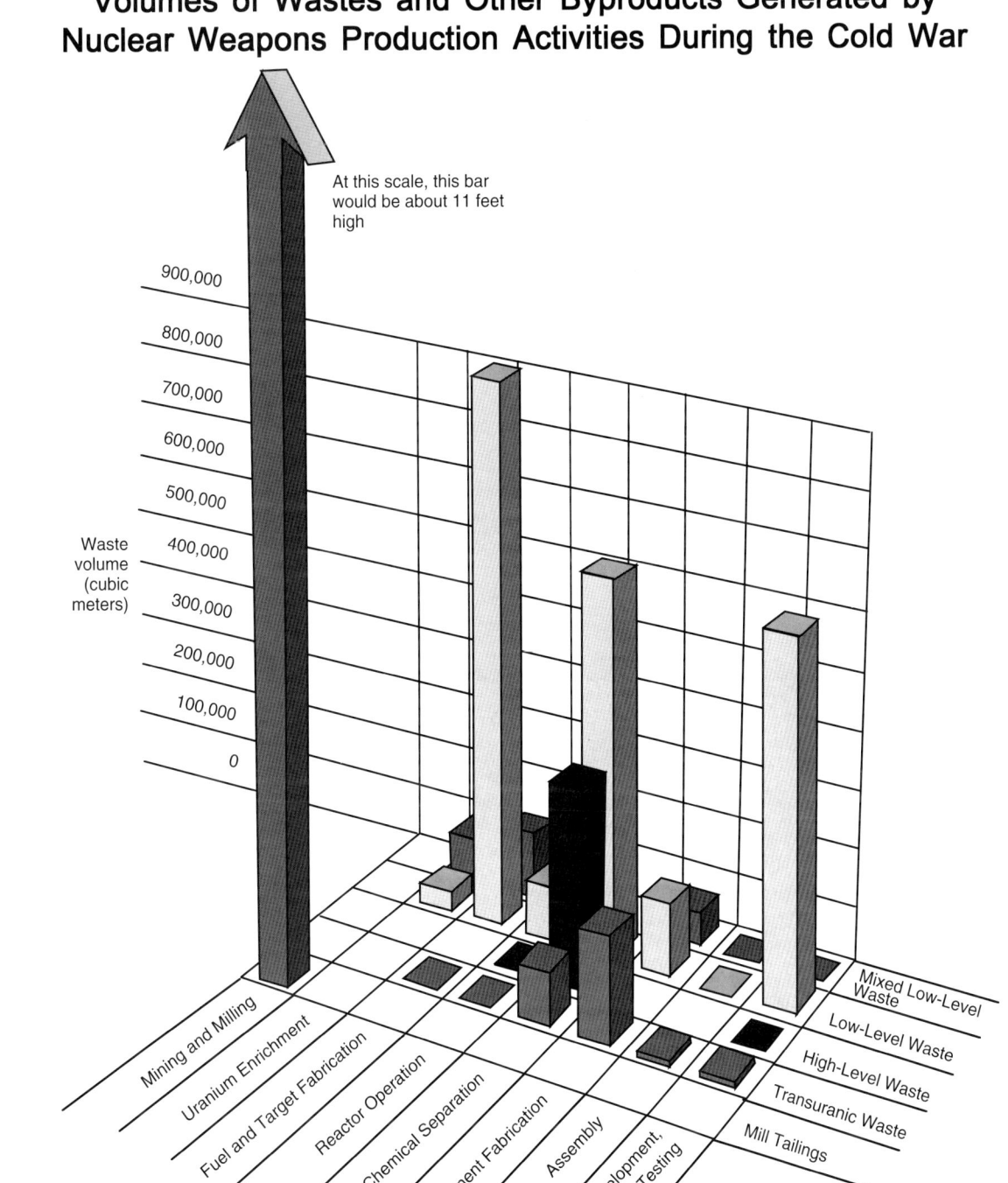

Volumes of Wastes and Other Byproducts Generated by Nuclear Weapons Production Activities During the Cold War

At this scale, this bar would be about 11 feet high

Note: The data are results from a report prepared for the Congress by the Office of Environmental Management.

Each step in the process of designing, producing, testing, and maintaining nuclear weapons produces wastes and other byproducts. Facilities across the United States have contributed to this process and generated a variety of wastes as a result. Knowing how much waste of what type has been generated by what steps in the process is critical for planning how to manage the wastes and possibly redesigning, for the future, the steps in the process to minimize the generation of these wastes and the attendant problems. This graph illustrates the volume of five types of waste and byproducts generated by nuclear weapons activities during the Cold War, with mill tailings accounting for about 96 percent of the total volume. Another method for measuring the waste is according to the amount of radioactivity contained in the various waste types (see page 32).

A cask for shipping spent fuel stands empty after its cargo of irradiated nuclear fuel has been deposited into the nearby spent-fuel pool for storage. A worker is completing decontamination of the cask so that it can be reused. The spent-fuel pool in the background holds 22 million gallons of water. *Idaho Chemical Processing Plant, Fuel Storage and Treatment Facility, Building 666, Idaho National Engineering Laboratory. March 17, 1994.*

Spent Nuclear Fuel

To produce plutonium and tritium for nuclear warheads, the United States operated 14 nuclear reactors. The first one started in 1944; the last one was shut down in 1988. During that time, most of the nuclear fuel rods and targets irradiated in the reactors were reprocessed to extract the plutonium as well as the leftover enriched uranium for reuse. The process produced liquid high-level waste, transuranic waste, low-level waste, and mixed waste.

During the Cold War, the Government stored its spent-fuel elements before reprocessing – and only as long as necessary for them to "cool off" by radioactive decay. Declining demand for plutonium and highly enriched uranium, however, steadily reduced the pace of reprocessing. When the Department announced the phaseout and eventual complete cessation of reprocessing in April 1992, it had accumulated approximately 2,700 metric tons of spent fuel in nearly 30 storage pools. About 99 percent of this spent fuel is stored at four sites: the Hanford Site in Washington, the Savannah River Site in South Carolina, the Idaho National Engineering Laboratory, and West Valley in New York. Most spent fuel is stored indoors, in pools under water that is cooled and filtered; some spent fuel is kept in dry storage.

The amount of spent fuel stored by the Energy Department is much smaller than the amounts stored by the commercial nuclear power industry, but Department of Energy fuel often presents greater safety problems. The commercial industry currently stores approximately 30,000 metric tons at more that 100 nuclear reactor sites around the United States; this is about 10 times the mass stored by the Energy Department. Unlike fuel for commercial nuclear reactors, however, most of the Department's spent fuel was designed to be reprocessed. Its cladding – the outer layer of zirconium metal – was not designed for long-term storage. As a result, some of the stored spent fuel has corroded, leading to a number of potential safety problems. Also, some of the Department's

Pool for the storage of spent fuel. This pool is 28 feet deep; 7 feet of water cover the top of the highly radioactive spent-fuel elements. Water cools the fuel and also acts as radiation shielding. *Receiving Basin for Offsite Fuel, Savannah River Site, South Carolina. January 7, 1994.*

spent fuel contains highly enriched uranium and thereby presents much greater security and safety concerns than commercial spent fuel.

The Department's challenge is to safely store this spent fuel for the years that will pass before a geologic repository is available for permanent disposal. Unfortunately, many existing storage facilities do not meet current commercial or government safety standards; some of them are nearly 50 years old. Some pools are unlined and do not have adequate provisions for the control of water chemistry, a situation that is likely to lead to corrosion and leakage. A lesser concern has been the potential for an inadvertent nuclear chain reaction (a so-called "criticality event") resulting from accidents in handling or storage.

Ninety-nine percent of government owned spent fuel is stored in four states: Washington, South Carolina, Idaho, and New York.

Corroding spent-fuel elements from the Hanford N Reactor are stored in an unlined concrete pool in the 105 K-East area. Corrosion releases radioactive materials to pool water, posing a hazard to workers. *Hanford Site, Washington.*

Straddle carrier for moving casks of spent fuel into dry storage. The Department of Energy is replacing underwater pool storage with these dry casks to increase safety and reduce costs. *Idaho National Engineering Laboratory. March 17, 1994.*

Reducing Risks from Spent-Fuel Storage

The Department of Energy has evaluated its facilities for spent-fuel storage and it is developing new storage methods and facilities. Material posing the highest risk is being moved out of inadequate facilities, repackaged and stabilized, and placed in more secure locations. For example, a spent-fuel storage pool at the Idaho National Engineering Laboratory is earthquake resistant, can retard corrosion by maintaining proper water chemistry, and has a leak-detection system. Spent fuel from other areas at the Laboratory is being consolidated there.

At the Hanford Site, radioactive sludge and spent fuel exist in an obsolete facility a few hundred yards from the Columbia River. In the past, one basin leaked millions of gallons of contaminated water into the ground. The spent fuel and sludge will be packed in new containers and moved away from the river to a modern storage facility. An environmental study is considering long-term dry storage of Hanford spent fuel. In the meantime, Hanford's fuel pools are being upgraded to minimize the potential for leaks and render them less susceptible to earthquake damage.

> *Spent nuclear fuel that poses the highest risks is being moved out of inadequate facilities, repackaged, stabilized, and placed in more secure locations.*

Providing dry aboveground storage for spent fuel in special casks is one possible alternative to underwater storage. *Spent Fuel Storage-Cask Testing Pad, Test Area North, Idaho National Engineering Laboratory. March 17, 1994.*

Options for the Long-Term Storage and Disposal of Spent Fuel

The Department completed in 1995 a comprehensive national environmental study to decide whether to leave the spent fuel at the sites where it is located or to consolidate it in a few regional locations or in one central place. The option selected was to store similar types of spent fuel together to optimize use of technical management expertise and in case some preparation of the spent fuel was required for long-term storage and disposal.

The Department is testing aboveground dry-cask storage designs for spent fuel that has cooled long enough in pools. Dry casks typically provide more reliable long-term storage than pools. Many commercial nuclear power plants already use this storage method. The Idaho Underground Dry Vault Storage Facility demonstrates a version of this method for the storage of spent fuel. One candidate for dry storage is the N Reactor spent fuel from the Hanford Site.

The current plan for the disposal of spent fuel– either as intact fuel elements or in some other form – is emplacement in a geologic repository mined deep in stable rock. There is widespread international agreement that this method of disposal can provide long-term isolation. Any spent fuel destined for geologic disposal will first have to be encapsulated in metal containers designed to meet regulations for performance in a repository. In some cases the spent fuel may require processing to prepare it for disposal or long-term storage. For example, damaged fuel may present too great a risk for storage. Also, spent fuel containing weapons grade, highly enriched uranium may require processing to avert potential security and criticality problems during storage or after disposal. The Department is considering new technologies for stabilizing spent fuel without reprocessing, which creates waste, contamination, radiation exposure and non-proliferation problems and is very costly.

A more detailed discussion of geologic disposal can be found on pages 45 and 46, although the repository described there is intended exclusively for transuranic waste.

The T Plant was the world's first reprocessing canyon. In 1944, it dissolved spent fuel from the Hanford B Reactor and chemically extracted the plutonium, which was then used to form the core of the Trinity and Nagasaki bombs. It continued reprocessing until 1956. Today, the plant is used to decontaminate equipment. *Hanford Site, Washington. July 11, 1994.*

High-Level Waste from Reprocessing

Irradiated fuel and target elements discharged from a production reactor contain a variety of intensely radioactive fission products (the lighter isotopes resulting from the splitting of uranium) mixed with the desired plutonium and uranium. During the Cold War, these fuel elements were dissolved in acid and chemically processed to separate the plutonium and uranium. The acids and chemicals from these operations are known as "high-level waste." Nearly all of the fission products resulting from irradiation are contained in this liquid high-level waste.

High-level waste is the most radioactive byproduct from reprocessing and contains most of the radioactivity originally found in the spent fuel. The intense radioactivity is caused by the relatively rapid decay of many fission products. As a result, it will generate one-tenth as much heat and radiation after 100 years, and it will have decayed to 1 one-thousandth of its original level in 300 years. The decay helps make the handling of the waste safer and easier. Nonetheless, the waste will require disposal and isolation from the environment for a very long time, essentially as long as spent fuel.

The liquid high-level waste resulting from reprocessing is stored in 243 large underground tanks in four states.

For reprocessing operations, five facilities were built at Hanford, two at the Savannah River Plant, and one at the Idaho National Engineering Laboratory. These buildings and their underground tanks for high-level waste are among the most radioactive places in the United States. Four of the Hanford canyons and one at the Savannah River Site were primarily devoted to plutonium extraction. Two others (the second canyon at Savannah River and the one in Idaho) were used for extracting highly enriched uranium from spent fuel. The fifth Hanford canyon was briefly used to recover uranium from high-level-waste tanks. In addition, a demonstration plant for reprocessing commercial spent fuel was built and operated briefly in West Valley, New York. The high-level waste from this plant is also the Department's responsibility.

Million-gallon double-walled carbon-steel tanks under construction at Hanford. These tanks are designed to contain high-level radioactive waste from plutonium-production operations. They will replace older single-walled tanks, many of which have leaked. The new tanks are designed to last for 50 years. By that time it is believed that a long-term solution for high-level-waste disposal will have been developed. *Hanford Site, Washington. November 16, 1984.*

The Department currently stores about 100 million gallons of high-level waste–enough to fill about 10,000 tanker trucks–the largest volume of waste in the Department's inventory. Most of this waste has been stored in 243 underground tanks in Washington, South Carolina, Idaho, and New York. The waste stored in these tanks contains a variety of radioactive liquids, solids, and sludges. Some of the liquid has been converted to a concentrated dry form. Because workers during much of the Cold War often filled these tanks without first sampling the waste and without recordkeeping to today's standards, the Department does not have complete knowledge of some waste characteristics. If high-level waste is inadequately managed, it can pose serious immediate as well as long-term risks.

The older Hanford tanks were designed for a useful life of 25 years. By 1973, 15 of the tanks had experienced significant leaks into nearby soil and ground water. Currently 67 tanks at Hanford are known or suspected of having leaked high-level waste into the surrounding soil. The three largest leaks released 115,000, 70,000, and 55,000 gallons of high-level waste.

Reducing Risks from High-Level Waste

In some tanks, radioactive decay and chemical reactions generate hydrogen gas or other compounds that can explode under certain conditions. While the Soviets experienced an explosion of high-level waste with serious public health consequences in 1957, such an accident is not likely in the United States because both the chemical constituents of the waste and the storage conditions differ. It is, however, important to understand the circumstances of the event to ensure that it does not occur in the United States. The Department has made a major effort in recent years to reduce the possibility of a waste-tank explosion at Hanford.

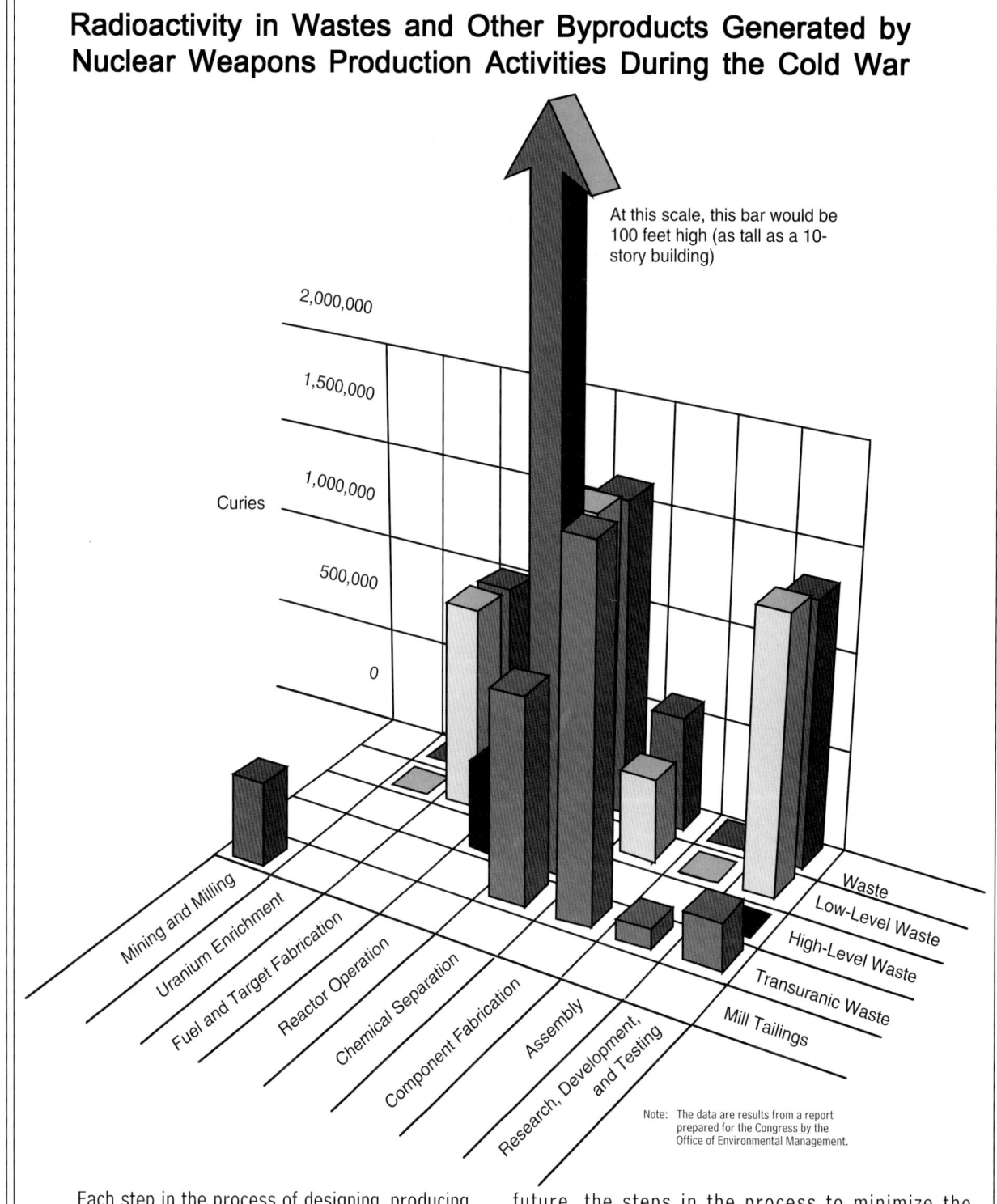

Each step in the process of designing, producing, testing, and maintaining nuclear weapons produces wastes and other byproducts. Facilities across the United States have contributed to this process and generated a variety of wastes as a result. Knowing how much waste of what type has been generated by what steps in the process is critical for planning how to manage the wastes and possibly redesigning, for the future, the steps in the process to minimize the generation of these wastes and the attendant problems. This graph illustrates the amount of radioactivity contained in five types of waste and byproducts generated by nuclear weapons activities during the Cold War, with high-level waste from chemical separation accounting for 99 percent of the radioactivity. Another method for measuring the waste is by volume. (see page 25).

Byproducts of the Cold War

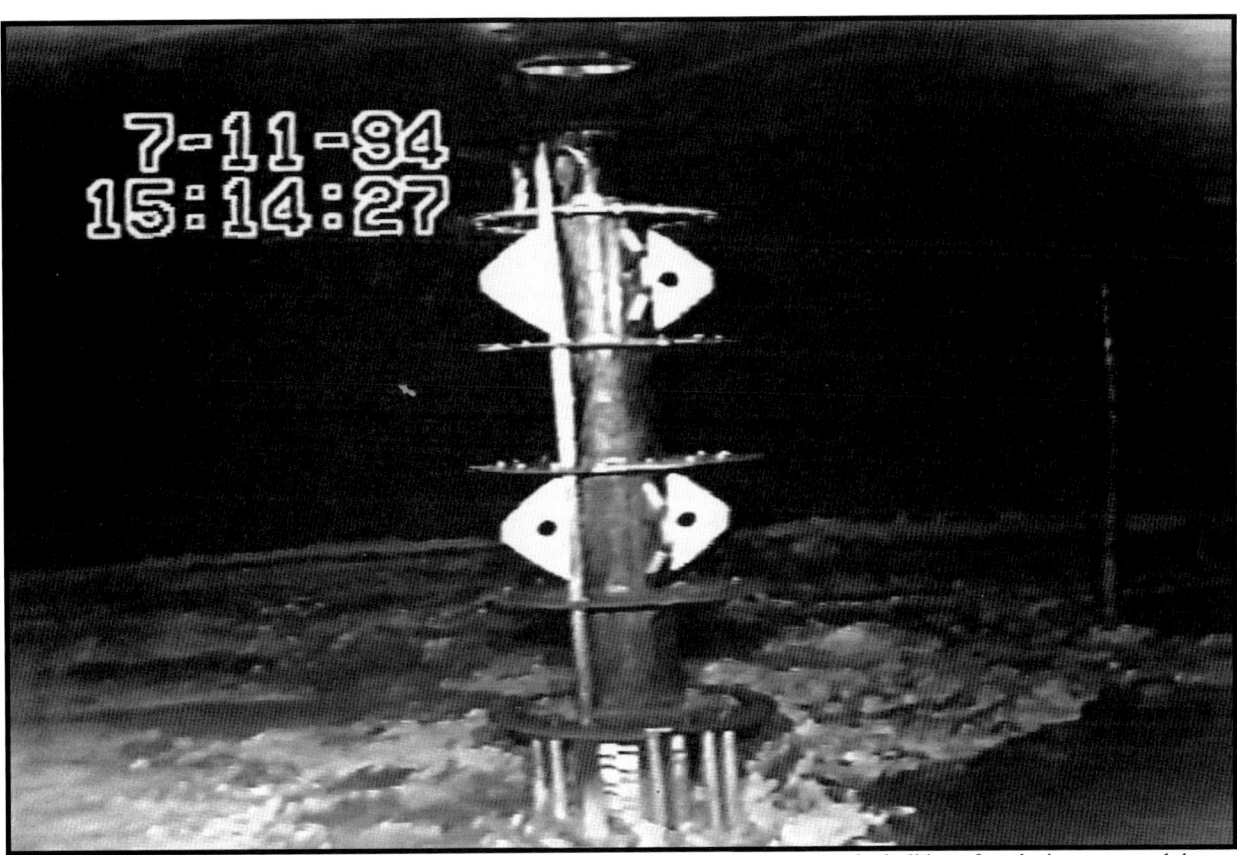

A mixing pump inside a storage tank slowly stirs high-level waste. This action prevents the buildup of explosive gases and thus minimizes the risk of an explosion. *Tank 241-SY-101, Hanford Site, Washington. July 11, 1994.*

Advanced robotics for cleanup of high-level waste are demonstrated by research scientist Jae Lew. The robotic manipulator in the distance is designed to break up and remove sludge and solidified waste inside a high-level-waste tank. This system is also used to develop, test, and evaluate a variety of methods for the retrieval of high-level waste. *337 Building, Hanford Site, Washington. July 11, 1994.*

A special mixing pump has been designed and installed at Hanford in the tank identified as having the highest risk of a hydrogen-gas explosion. Hydrogen had accumulated in the solids in the lower part of the tank, where it periodically "burped" up to the surface and into the tank's airspace. A spark could have caused an explosion, releasing high-level waste to the environment. The mixing pump circulates the waste in the tank, allowing hydrogen to escape at regular intervals and in safe concentrations through a filtered ventilation system, virtually eliminating the threat of explosion. A backup pump has been built and is ready to be installed if needed. Mixing pumps may also be installed in other tanks.

Another chronic problem at Hanford is that most of the original storage tanks for high-level waste were single-walled tanks made of carbon steel. The carbon steel corroded, and no provision had been made to contain material that leaked out of the tanks.

To help correct this problem, 28 new double-walled tanks of carbon steel and concrete were constructed in the 1980s with a life expectancy of about 50 years. Most of the free-standing liquids from the single-shell tanks has been transferred into the new tanks.

Some of the stored high-level waste is in a solid "saltcake" form. At present, this solid waste cannot be removed from its storage tanks without first dissolving it with water. The Department is designing advanced robotics equipment, controlled by operators from a safe distance, that will be capable breaking up and extracting this material.

Stabilizing High-Level Waste: Preparing for Disposal

Even after tens of thousands of years, high-level waste will remain radioactive. The Department of Energy is thus charged with ensuring that these materials are isolated from people and the environment for a very long time. In preparation for long-term disposal, the Department is developing ways to put the most radioactive byproducts of nuclear weapons production into more stable forms.

In a plant for calcining high-level waste, manager Judy Burton monitors the controls of the fluidized bed used to heat liquid high-level waste and convert it to powder. This reduces the volume of the waste by up to eight times. *Idaho National Engineering Laboratory. March 17, 1994.*

In Idaho, workers have converted much of the liquid high-level waste to a dry concentrated powder and stored it in bins, ready for final treatment in preparation for disposal.

Progress in Idaho

The Idaho National Engineering Laboratory has operated a calcining facility that uses heat to convert large quantities of liquid high-level waste into a dry powder for storage. The calcined waste occupies up to eight times less volume and is more stable than the liquid waste. The Department is upgrading the calcining plant and support facilities to meet current safety and environmental requirements.

After calcining, the powdered waste is stored in steel silos housed inside cylindrical concrete bins several feet thick. The vaults are engineered to contain the waste and to provide passive cooling. Direct human contact with the waste would be dangerous, and the dry waste could be dispersed easily. The Department is assessing which technology would be most suitable for converting the material into a more stable form for disposal in a permanent repository.

Despite this success in stabilizing waste, some high-level waste in Idaho remains in liquid form. Its high sodium content prevents calcining without significant dilution or treatment. Engineers are now developing methods to calcine the remaining liquids.

At other sites where reprocessing created this type of waste, workers are taking a different approach, which skips this intermediate step.

This storage bin for calcined high-level waste is made of reinforced concrete and steel. Inside its 4-foot-thick walls are stainless-steel silos containing up to 55,000 cubic feet of high-level waste in powdered form. There are seven bins like this in Idaho, and they are engineered to provide safe storage for 500 years. *Idaho National Engineering Laboratory. March 17, 1994.*

The Department is upgrading the calcining plant and support facilities to meet current safety and environmental requirements.

Converting Waste to Glass in South Carolina, New York, and Washington

At the three other reprocessing sites, high-level liquid acidic wastes were neutralized for storage in carbon steel tanks. The resulting liquids, sludge, and saltcake will be mixed with molten glass, and poured into metal cylinders. Similar processes are already being used in Europe.

This method, called "vitrification," poses a number of technical challenges. Any plant that processes high-level waste must be shielded and operated by remote control. In addition, some of the waste needs to be chemically treated to prepare it for vitrification. The process must be controlled carefully to avoid tank corrosion or the generation of dangerous gases. The waste to be treated has a variety of chemical forms that might prove difficult to blend with molten glass.

The Department has constructed two of the world's most modern radioactive-waste vitrification facilities and has completed major testing prior to waste vitrification. At the South Carolina plant, workers produced more than 70 canisters of test glass in 1995. In addition, chemical treatment of wastes in preparation for vitrification was completed. The Savannah River Site in South Carolina plans 20 years of operation to vitrify existing high-level wastes. The other facility is a smaller plant at West Valley, New York, near Buffalo. The backlog of high-level waste at this plant will take several years to vitrify.

Vitrified waste will be poured into stainless-steel canisters that will be placed in a storage facility. In this form, the waste will cost much less to store and monitor than liquid waste. Once a geologic repository is ready, the canisters will be transported there for permanent disposal. If the Yucca Mountain, Nevada site that is currently the subject of characterization studies proves suitable, the Department expects to begin sending its high-level waste there by 2010.

The Department of Energy is preparing to stabilize the most radioactive byproducts of nuclear weapons production for long-term storage and disposal.

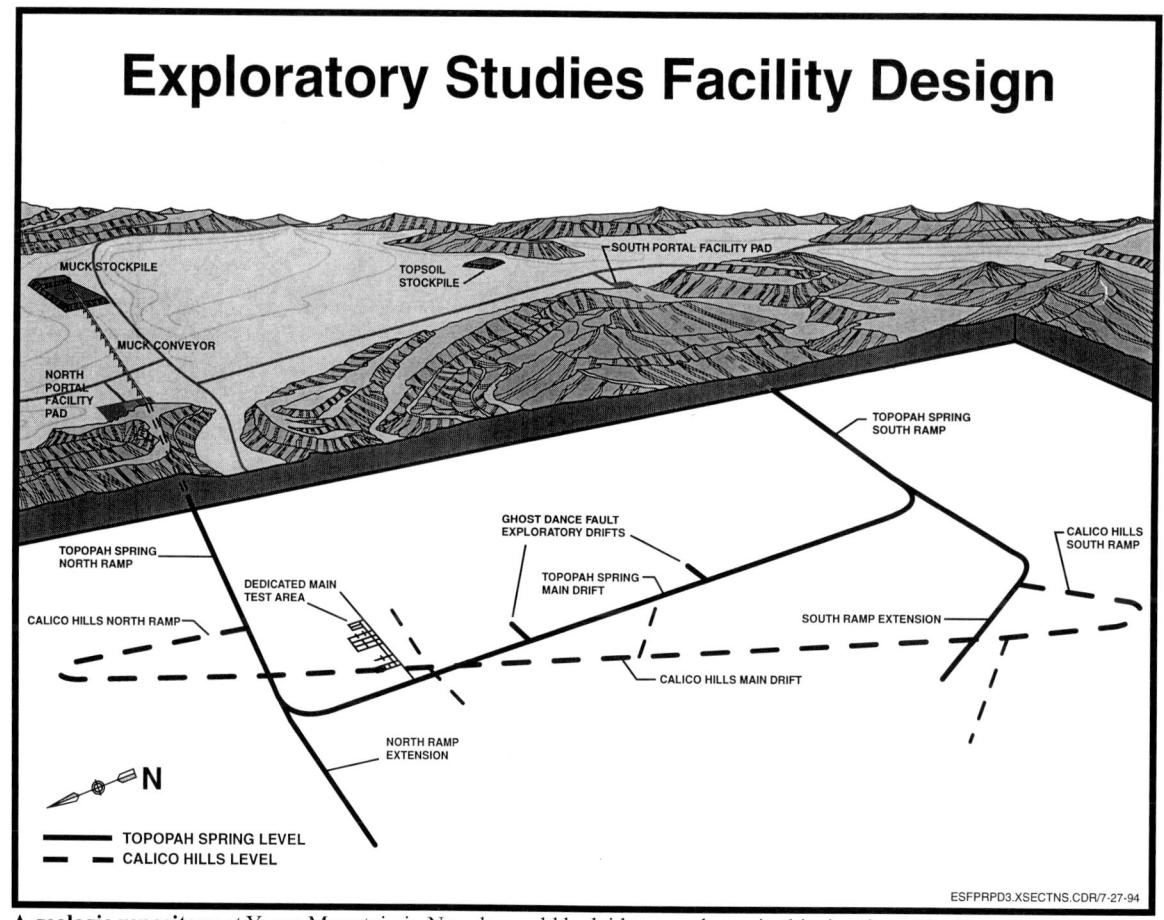

A geologic repository at Yucca Mountain in Nevada would be laid out as shown in this drawing. The Yucca Mountain site has been studied for over 10 years to determine whether it is suitable for a repository. If the site is found to be suitable, the Department expects to start sending its waste to this geologic repository by the year 2010.

Byproducts of the Cold War

This vitrification plant for high-level waste is 120 yards long and encompasses 5 million cubic feet. It contains 69,000 cubic yards of concrete with 13,000 tons of reinforcing steel and 320,000 feet of electrical cable. It is designed to turn high-level waste into glass logs by pouring a mixture of waste and borosilicate glass into stainless-steel canisters, which are then sealed and stored. Workers completed major testing in 1995, including producing more than 70 canisters of test glass. *Defense Waste Processing Facility, Savannah River Site, South Carolina. January 7, 1994.*

These stainless-steel canisters weigh 1,100 pounds each. When full of vitrified waste, they will weigh 3,700 pounds each and will be extremely radioactive. *Defense Waste Processing Facility, Savannah River Site, South Carolina. June 15, 1993.*

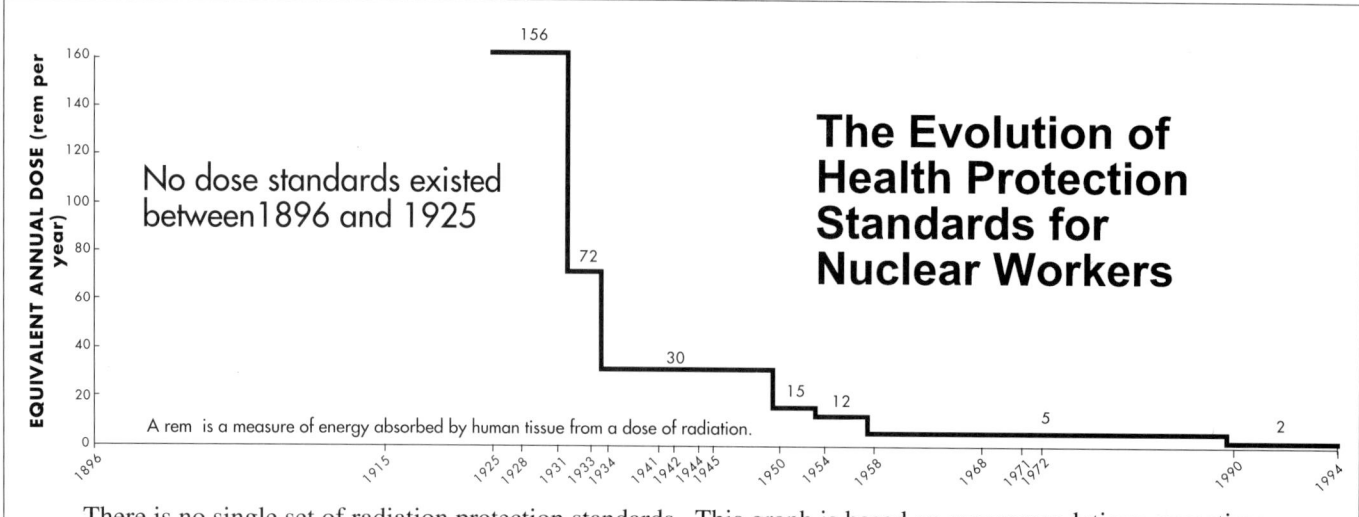

There is no single set of radiation protection standards. This graph is based on recommendations, sometimes different, published by U.S. and international groups concerned with radiation protection. They have been translated into a single, consistent set of numbers and measurement units for the purpose of this summary.

1896 Henri Becquerel discovers radiation. First radiation injuries are reported, but no protection standards exist.

1915 Protection standards describing "safe practices" for handling radium and X-ray machines are published in Sweden and Germany. Radiologists are advised to stay as far away from their equipment as possible, to handle radium vials with tongs, and to work no more than 35 hours a week. The U.S. and Britain soon follow suit, but no dose limits are set because measurement techniques and units do not yet exist.

1925 Swedish and German scientists publish estimates of "tolerance doses," the amount of radiation a person is thought to absorb without harm. Based on the amount of radiation that would burn skin, the tolerance dose is initially estimated to be the equivalent of about 156 rem per year (over 45 times the current standard), although the estimates vary widely.

1928 The first internationally accepted X-ray protection standard, 1 one-hundredth of the amount that burns skin per month, is accepted at an international congress.

1931 The tolerance dose is standardized at 6 rem per month (72 rem per year).

1933 The genetic effects of radiation on fruit flies are studied by German scientist A. Mueller. He learned that radiation caused genetic mutations.

1934 First international radiation safety standards based on measurements of damage to human tissue are published in Zurich by the International Commission on X-Ray and Radium Protection. Workers are allowed up to 0.1 rem per day (30 rem per year).

1941 Recommended tolerance for ingested radium is initially set at 1 ten-millionth of a curie per person by the National Commission on Radiation Protection. This recommendation is based on studies of radium-watch-dial painters.

1942 The Manhattan Project begins. The 1934 radiation exposure standards of 30 rem per year are accepted by the University of Chicago's Metallurgical Laboratory after experimental verification. The "tolerance" concept is discarded in favor of the "maximum permissible exposure."

1944 The initial tolerance limit for plutonium inhalation is set at 5 millionths of a gram per person by the Manhattan Project's radiation protection laboratory.

1945 The first atomic bombs are produced, tested, and used. Weighting factors for the different types of radiation are introduced to account for their different health effects. The plutonium tolerance limit is lowered to 1 millionth of a gram per person.

1950 Scientists discard the idea of a "maximum permissible exposure," recognizing that any amount of radiation may be dangerous. Radiation protection scientists recommend that exposure be "as low as reasonably achievable." Concern over latent cancer, life shortening, and genetic damage also causes standards to be halved: 0.3 rem per week (15 rem per year).

1954 A quarterly limit of 3 rem per 13 weeks (12 rem per year) is introduced by the U.S. National Bureau of Standards to allow more flexibility in exposure patterns. Workers are still allowed 0.3 rem per week up to this limit.

1958 In response to a study by the National Academy of Sciences of the genetic effects of radiation, a new dose limit is introduced, using a formula that allows workers to receive 5 rem per year after the age of 18. Annual doses are allowed to exceed this level up to 3 rem per 13 weeks (12 rem per year). To protect the gene pool, a lower standard of 0.5 rem per year is set for the general public.

1968 The Federal Government updates its protection standard to the 5 rem per year recommended in 1958. This standard has not been changed since.

1971 Radiation protection standard is restated by the National Committee on Radiation Protection but not really changed: 3 rem per 13 weeks in the past, 5 rem per year in the future. By including exposure from internal radiation ("body burden"), the standard is effectively lowered by a significant amount.

1972 The National Academy of Sciences publishes its first study of the health effects of radiation since 1956. The report, Biological Effects of Ionizing Radiation I (BEIR I) becomes the first of a series.

1990 The National Academy of Sciences BEIR V report asserts that radiation is almost nine times as damaging as estimated in BEIR I. Annual doses may no longer exceed 5 rem per year. The International Commission on Radiation Protection recommends that an average dose of 1 or 2 rem per year not be exceeded.

Radiation and Human Health

Before 1896, scientists believed that atoms were immutable and eternal. The discovery of radiation changed this view forever. Since its discovery, scientists have studied radiation intensely. Its potential for commercial and medical benefits, and its health risks, became quickly apparent. In comparison with many nonradioactive chemicals, radiation is easy to detect and measure, and hundreds of studies have quantified its effects on living organisms. Nonetheless, it is not possible to predict its exact effects on a specific person. There is no doubt that high levels of radiation cause serious health damage. The precise effects of low-level radiation continue to be controversial.

What Is Radiation?

Radiation is energy emitted in the form of particles or waves. Radioactive materials like radium are naturally unstable and spontaneously emit radiation as they "decay" to stable forms. Although the term "radiation" includes microwaves, radiowaves, and visible light, we are referring to the high energy form called "ionizing" radiation (i.e., strong enough to break apart molecules), which produces energy that can be useful, but can also damage living tissue.

Kinds of Radiation

There are four major types of radiation:

Alpha particles are heavy particles, consisting of two neutrons and two protons. Because the particles are slow moving as well as heavy, alpha radiation can be blocked by a sheet of paper. However, once an alpha emitter is in living tissue, it can cause substantial damage.

Beta particles consist of single electrons. They are moderately penetrating and can cause skin burns from external exposure, but can be blocked by a sheet of plywood.

Gamma rays are high-energy electromagnetic rays similar to X-rays. They are highly penetrating and several inches of lead or several feet of concrete are necessary to shield against gamma rays.

Neutrons are particles that can be both penetrating and very damaging to living tissue, depending on their energy and dose rate.

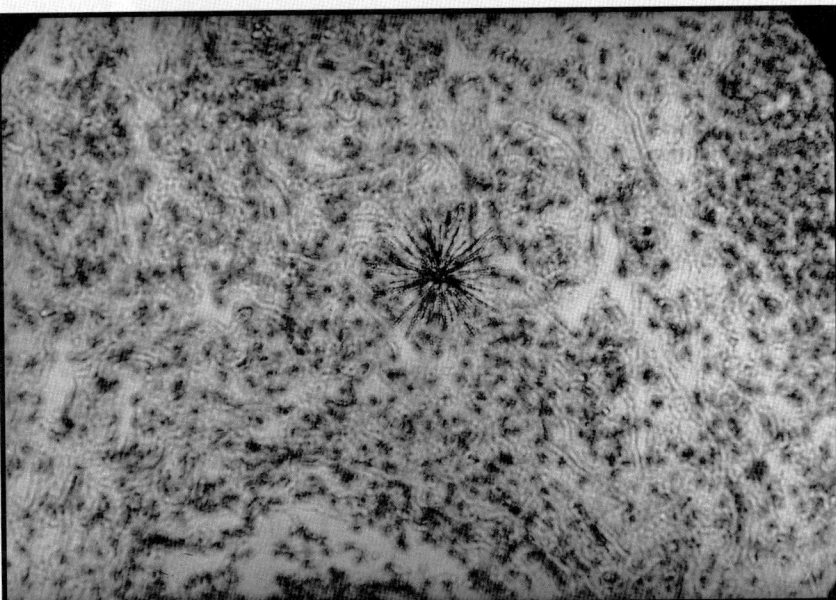

Particle of plutonium in lung tissue. The black star in the middle of this picture shows tracks made by alpha rays emittted from a particle of plutonium in the lung tissue of an ape. Alpha rays do not travel far, but once inside the body they can penetrate the more than 10,000 cells within their range. *Magnification 500 times. Lawrence Radiation Laboratory, Berkeley, California. September 20, 1982.*

Measuring Radiation

One way to measure radiation is at its source. This is done by monitoring the rate at which the atoms in a radioactive element disintegrate. This mechanical measurement uses the "curie" as its basic unit, 1 curie being 37 billion atomic disintegrations in 1 second.

A different way is to calculate radiation energy at its point of impact in the body. This is the health-based approach. Its basic unit of measurement is the rem (roentgen-equivalent-man), and it is based on assumptions about the actual damage or accumulation of radioactivity in body parts, such as bones or lungs. These assumptions result in some uncertainty, but this approach allows more meaningful measurements than measuring energy levels from a source. Since a single radiation dose has different effects on different body organs, it is not easy to predict what effect a given dose will have on a person's health.

Half-Life

The less stable an atom, the more rapidly it breaks down and the shorter its half-life—the time required for half of the original atoms to decay. During a second half-life, half the remaining atoms, or one-quarter of the starting number, will decay, and so on. The half-lives of various isotopes range from fractions of a second to billions of years.

How Can Radiation Cause Damage?

In living organisms, the chemical changes induced by high doses of radiation can lead to serious illness or death. At lower doses, radiation can damage DNA, sometimes leading to cancer or genetic mutations. Even the natural background radiation level (which depends on geographic location, altitude, and other factors) imposes some risk of illness. An estimated 82 percent of the average radiation exposure received by people in the United States comes from natural sources.

Understanding Radiation Hazards

Measuring a substance's radioactivity is only the first step toward understanding its potential hazards to living organisms. Other important factors include:

<u>Type of radioactivity</u>. Some radiation, such as alpha particles, can cause chemical changes at short range. Other kinds, such as neutrons, can be harmful from distant external sources.

<u>Chemical stability</u>. Radioactive substances that can burn or otherwise react are more susceptible to being dispersed into the environment. For instance, some forms of plutonium can spontaneously ignite if exposed to air.

<u>Biological uptake</u>. Radioactive elements incorporated into organisms are more harmful than those that pass through quickly. Many radioactive elements are readily absorbed into bone or other tissues. Radioactive iodine is concentrated in the thyroid, while radium and strontium are deposited in bone. Insoluble particles like plutonium oxide can remain in lung tissue indefinitely.

<u>Dose and dose rate</u>. Dose rate is the amount of radiation received in a given time period, such as rem per day. In general, the risks of adverse health effects are higher when exposure is spread over a long period than when the same dose is received at one time.

<u>Dose location</u>. Some kinds of living tissue are more sensitive to radiation than others.

The combined effect of the above factors makes the risk posed by even a simple radiation exposure difficult to estimate. Real-world wastes from nuclear weapons production often contain many different radioactive constituents—along with various chemicals—introducing even more uncertainty. However, the hazards can be better defined by considering the particular types of radiation emitted by each radioactive element and by modeling likely pathways of exposure.

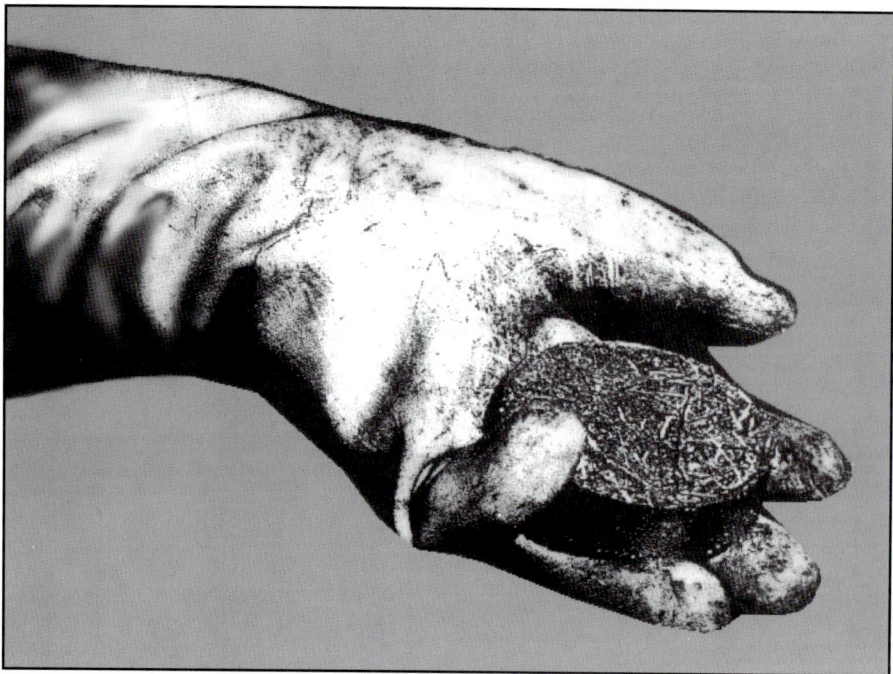

Plutonium metal puck. Plutonium must be handled and stored in small quantities like this to prevent it from spontaneously starting a nuclear chain reaction. *Rocky Flats, Colorado.*

The Plutonium Problem

Plutonium can be dangerous even in extremely small quantities, particularly if it is inhaled as a dust. Finely divided plutonium metal may ignite spontaneously if it is exposed to air above certain temperatures. Therefore, extraordinary precautions are required when handling it. The facilities that processed plutonium chemically and metallurgically or made the plutonium into high-precision warhead components are structurally similar to electronics industry "clean rooms" or research labs for the study of virulent diseases. Plutonium-production operations are enclosed in gloveboxes, which are filled with a dry inert gas or air at pressures lower than normal room air pressure. That way, if a leak develops, contamination will not flow outward.

The X-Y Retriever Room at the Rocky Flats Plant contains plutonium in many forms. During normal operations, plutonium in this room was recycled for warhead production. Today the room is used to store surplus plutonium. Here workers are making repairs to a hoisting mechanism. The hoist is used to lift the plutonium from the storage cans on the floor. *Building 707, Rocky Flats Plant, Colorado. November 29, 1988.*

Workers in these plants wear special anticontamination coveralls, rubber shoe covers, and two layers of surgical gloves – and, when necessary, respirators or "moon suits." Disposable coveralls, gloves, and shoe covers become radioactive waste after use on each work shift, every workday. Technicians carefully scan all employees with radiation detectors when they enter or leave certain areas to ensure that they have not been contaminated. The process is time-consuming but necessary to ensure safety.

To prevent diversion by terrorists, plutonium requires constant protection against theft. To further complicate matters, it must be handled carefully to avoid putting more than a few kilograms of it in close proximity. This must be done to prevent a burst of radiation known as a "criticality event." An inadvertent criticality event would not cause a nuclear explosion, but it would release intense radiation that can penetrate the shielding used in plutonium operations, and the radiation could be lethal to nearby workers.

Plutonium Residues and Scraps

There are many steps in the manufacture of plutonium parts for nuclear weapons. The sudden shutdowns of plants that did this work, including the Rocky Flats Plant in Colorado, the Hanford Site in Washington, and the Savannah River Site in South Carolina, stranded 26 tons of plutonium in various intermediate steps. The plutonium is in a wide variety of forms, from plutonium dissolved in acid to rough pieces of metal to nearly finished weapons parts. Scraps of metal and chemicals that contain enough plutonium to be worth recovering were stored in drums and cans. Unknown amounts of plutonium have collected on the surfaces of ventilation ducts, air filters, and gloveboxes.

The safe management of plutonium requires vigilance and caution under the best circumstances. The complexity of conditions in weapons plants presents an even greater challenge. Radioactivity from plutonium, some of it dissolved in corrosive acids, is slowly destroying the plastic bags and bottles that contain it. Flammable hydrogen gas is accumulating inside some of the sealed cans, drums, and bottles that clutter aisles and fill the gloveboxes. Bulging and ruptured containers have already been found in several places. Some of the plutonium is in a flammable form. In some cases, plutonium may be accumulating on the bottoms of tanks, where enough of it could result in a criticality event.

Some barrels of residues from plutonium operations are stored in drums in the building in which they were processed. *Building 776/777, Rocky Flats Plant, Colorado. December 20, 1993.*

Not only must plutonium be constantly inspected, guarded, and accounted for, but the buildings that house it also must be maintained. Ventilation systems and air filters must work continuously and fire and radiation alarms must be tested regularly. Men and women who work at the plants are at risk. Although they are less likely, severe accidents could endanger the nearby public and contaminate the environment.

These problems are among the Department's top priorities. All of the most urgent plutonium problems are now being addressed. Some of the ultimate solutions will take years to implement, but the work has begun. Workers at the Rocky Flats plant have been emptying bottles, draining tanks and pipes, and solidifying the liquids they remove. Pipes are already shrink-wrapped so they will not leak. New drains are being installed where needed, since almost half the liquids in the pipes and tanks cannot be removed otherwise. This work must be thoroughly planned and carefully executed. Most of the liquid plutonium at Rocky Flats will be solidified within 2 to 3 years.

In some cases, entire plants may have to be restarted to clean them out. For example, at the Savannah River Site's two chemical separation plants, more than 95,000 gallons of liquid containing dissolved plutonium have sat in tanks for several years. The Department began processing these hazardous solutions to stabilize them in 1995. Other unstable nuclear materials will be stabilized using a pilot-scale vitrification facility at the Savannah River Site.

Plutonium Metal in Storage

There are also problems with the plutonium-metal "pucks," "buttons," and other solid forms of plutonium metal kept in the storage vaults. This plutonium was stored in metal containers overpacked with plastic bags, and the bags were then sealed. In some cases, however, there are no exact records of what is contained in the sealed packages. Furthermore, the plutonium "rusts" into a powder when exposed to air. This powder can burn, and it could be inhaled by workers who must handle it. To eliminate these problems, the containers are being opened so that the plutonium dust can be brushed off and "roasted" in a special oven, thereby converting it to a more stable form for storage. The metal and powder are then repackaged separately without plastic to prevent the problem from recurring.

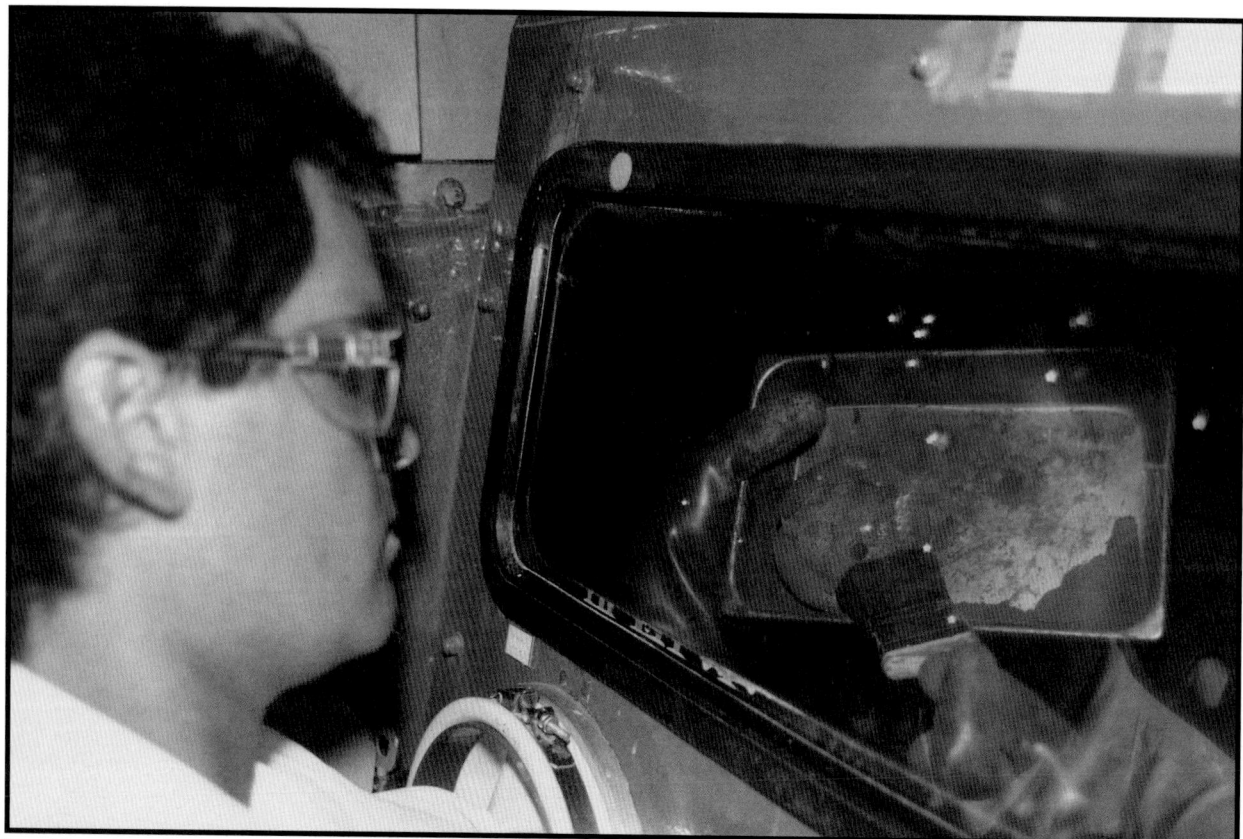

Brushing plutonium to remove oxidized portions is conducted inside gloveboxes. Scott Sterkel, a worker at Rocky Flats, carries out brushing on a plutonium button. The powder that is brushed off will be roasted to convert it to a more stable form. *Rocky Flats, Colorado. July 11, 1994.*

This high-security fence at Hanford's Plutonium-Uranium Extraction (PUREX) Plant was designed to safeguard strategic nuclear materials. Currently, this guardhouse, once staffed with guards, is used as an entry point for employees. PUREX operations ceased in 1990. *Hanford Site, Washington State. July 11, 1994.*

Informed Debate About Disposition

The United States produced and extracted more than 100 metric tons of plutonium for nuclear weapons during the Cold War; if the plutonium is not in operational warheads, it is currently stored at facilities across the country. The Department began thinking about switching from plutonium production to long-term storage and disposition even before the fall of the Soviet Union and the declassification of United States stockpile data. In February 1988, then Secretary of Energy John Herrington told a Congressional subcommittee that we were "awash in plutonium." In 1989, a National Academy of Sciences panel, using classified data, concluded that additional plutonium production was unnecessary. Now, however, the plutonium surplus continues to increase as each day more plutonium is removed from dismantled weapons at the Pantex Plant in the Texas panhandle and stored in World War II-era bunkers, at a rate of about 2,000 warheads per year.

The fate of all U.S. surplus plutonium must be determined publicly. Meaningful decisions about plutonium disposition can only be made through informed public debate, which has only recently begun with the release of vital information. For example, until Secretary of Energy Hazel R. O'Leary declassified plutonium stockpile information in December 1993, the public did not know how much plutonium the United States had produced (approximately 100 metric tons).

Scientists, engineers, policymakers, arms-control specialists, economists, and others are debating the fate of surplus plutonium in the United States. One fundamental question is whether the Department's plutonium is an asset or a liability. The United States spent billions of dollars to produce the plutonium it now possesses. Some argue we should recover this investment by fueling nuclear power plants with plutonium. Proposals have been made to fuel a new tritium-production reactor with it. Others contend this would be uneconomical, and we should find the safest, fastest, cheapest way to make it unusable for nuclear weaponry. One proposal is to vitrify it, just as is planned for high-level waste. Disposal suggestions have included deep geologic repositories, deep boreholes, and disposal in the ocean beneath the seabed. This issue is under intense study within the executive and legislative branches.

Drums of transuranic waste in interim storage in Idaho inside a tension-support structure. The waste in these drums will be disposed of at the Waste Isolation Pilot Plant (WIPP) if the repository meets all regulatory requirements. *WIPP Certification Station, Stored Waste Examination Pilot Plant, Idaho National Engineering Laboratory. March 17, 1994.*

The fate of surplus plutonium will be determined by addressing issues related to international security as well as environmental protection. Whatever decision is made in the United States will affect similar decisions being considered in other countries. One result could be smaller stockpiles of nuclear weapons material throughout the world. For example, using plutonium in a reactor or blending it with high-level waste could render it as inaccessible as if it were in spent nuclear fuel, which was the standard suggested by the National Academy of Sciences in a recent study.

While final decisions are being made, new technologies are needed to stabilize plutonium quickly without creating more radioactive waste than necessary. The Department has already developed two new technologies for this purpose at Hanford.

Another fundamental question revolves around the definition of plutonium wastes. Any material for which the cost of recovering the plutonium it contained was less than the cost of producing new plutonium was not previously considered waste. This definition is no longer appropriate after the end of plutonium production era.

Transuranic Waste

Nearly everything involved in plutonium processing becomes contaminated and must be contained and monitored indefinitely. Generally, such waste is called "transuranic" waste. Technically, this includes any material containing significant quantities of plutonium, americium, or other elements whose atomic weights exceed those of uranium. Transuranic waste can include everything from chemicals used in plutonium metallurgy to used air filters, gloves, clothing, tools, piping, and contaminated soils.

Accidents as well as normal operations have generated transuranic waste. The Rocky Flats Plant experienced numerous small fires in its production lines, and two major fires, in 1957 and 1969. Firefighting and subsequent decontamination efforts generated thousands of drums of waste, much of which was shipped to the Idaho National Engineering Laboratory for storage. Portions of the buildings are being decontaminated, and machinery and other wastes are being compacted and packaged for storage. Other problems, such as accidental releases of plutonium solutions, have rendered entire rooms in some buildings unusable.

Throughout the nuclear weapons complex, the transuranic waste inventory in storage totals about 100,000 cubic meters, or the rough equivalent of half a million 55-gallon drums. As in the case of spent fuel and high-level waste, much of the transuranic material was placed in temporary storage under the assumption that a permanent repository would soon become available. In the meantime, some containers have corroded, requiring costly cleanup, repackaging, and relocation.

Progress in Managing Transuranic Waste

In recent years, the Department of Energy has made a major effort at consolidating, repackaging, monitoring, and sheltering its transuranic waste. Transuranic waste has not always been stored with adequate safety. For example, thousands of drums have been exposed to the elements, risking corrosion and leaks. These are now being stored on concrete or asphalt pads under weather-resistant structures. Furthermore, much of the transuranic waste remains in earth-covered berms, which were expected to be needed for only a few years until a permanent disposal site became available. New storage facilities for this waste are being built, and drums that are corroding or leaking will be overpacked in clean metal containers. These interim steps will ensure safe storage until disposal in a geologic repository can begin.

Permanent Disposal

The long-lived radioactivity of plutonium, combined with the hazards if it is released even in small quantities, requires that transuranic waste be permanently isolated.

The Department of Energy has decided that deep underground disposal in a geologic repository is the best solution in terms of safety, cost, and practicality. This decision is based on recommendations by the National Academy of Sciences, many years of geologic investigations and experiments, and environmental studies. Waste in the proper forms and configurations, if emplaced in stable geologic formations, should be isolated with a high degree of confidence for tens of thousands of years.

Scientists in many countries agree that a geologic repository must be located in a rock formation with certain specific properties. For example, there must be evidence that the formation has been stable for millions of years; the rock

Demonstration models of special casks for shipping transuranic waste show how transuranic wastes will be trucked cross-country to the Waste Isolation Pilot Plant. Each of these TRUPACT-II (Transuranic Package Transporter) casks can hold fourteen 55-gallon drums. A window in the center cask displays mock waste drums cut open to reveal typical constituents of transuranic waste. This "roadshow" flatbed unit is used for public education and for training emergency-response teams along planned waste-shipment routes. *Waste Isolation Pilot Plant, near Carlsbad, New Mexico. February 25, 1994.*

must be free of circulating ground water; and the site should be located in an area with little potential for frequent and severe earthquakes or volcanic eruptions. In addition, the rock formation should be sufficiently deep beneath the surface and thick enough to allow the excavation of a repository with sufficient buffers of the same rock both above and below it. Also desirable is the absence of valuable natural resources which might attract inadvertent human intrusion into the repository in the distant future.

In the mid-1970s, the Department identified a site in southeastern New Mexico, near Carlsbad, as a promising candidate. The chosen rock formation was a thick layer of rock salt that had been deposited some 200 million years ago. The repository was to be excavated 2,150 feet below the surface. After environmental studies were completed in 1979, the Congress authorized the Department to build the repository, called the Waste Isolation Pilot Plant (WIPP). Large rooms have been excavated in the salt, and they are connected to the surface by several shafts to provide ventilation and to move excavated rock and waste containers. Surface facilities to receive the waste, inspect it, and move it underground have been built and equipped.

> *To create the WIPP, the Department excavated tunnels 2,150 feet deep in a thick layer of rock salt deposited 200 million years ago.*

Many experiments have been completed or are underway at the WIPP site to provide a better understanding of how the salt in the repository will behave and how waste materials will interact with the underground environment. No wastes have been taken to the site yet.

A vital part of the process for determining the suitability of the WIPP for disposal is providing opportunities for public involvement. Citizens groups, Native American Tribes, State and Federal agencies, and an independent technical review panel have been involved in a process to determine whether the WIPP can provide the required isolation for at least 10,000 years. The final decision will be made by the Environmental Protection Agency, which will assess the expected performance of the WIPP to determine whether it will meet environmental standards for the disposal of transuranic waste. If the decision is favorable, shipments of waste could begin in 1998.

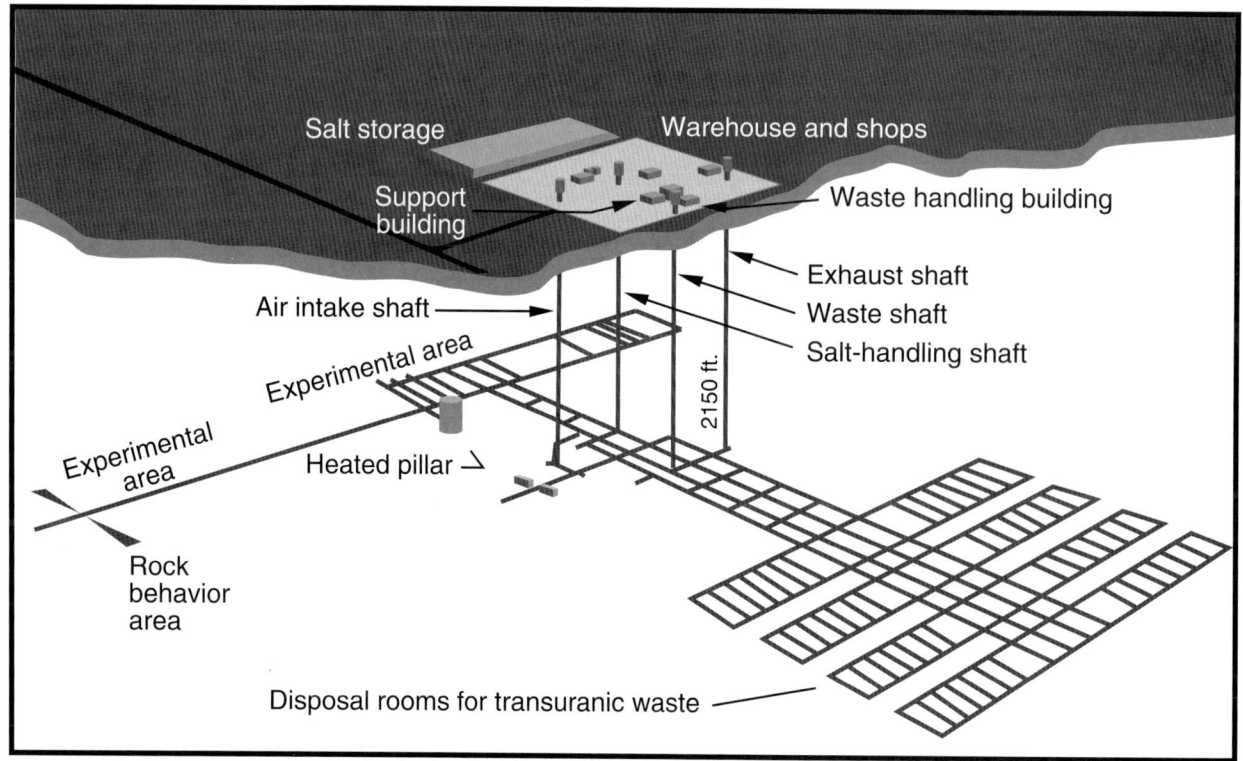

A simplified layout of the Waste Isolation Pilot Plant, showing the surface facilities, the four shafts, the underground areas in which experiments are conducted, and the underground rooms in which transuranic waste will be disposed of if disposal is approved.

Byproducts of the Cold War

Emergency exhaust airway at the WIPP. Should an accidental release of radiation occur within the WIPP's system of underground chambers, exhaust ventilation air would be diverted through a bank of filters to clean the air before it is released through this duct. *Waste Isolation Pilot Plant, Carlsbad, New Mexico. February 25, 1994.*

This underground waste-disposal room, excavated in 1986, was the first of 56 chambers to be excavated at the WIPP. It is 300 feet long, 33 feet wide, and 13 feet tall and could hold six thousand 55-gallon drums of transuranic waste. It lies 2,150 feet below the surface of the earth. *Room 1 of Panel 1, Waste Isolation Pilot Plant, near Carlsbad, New Mexico. February 25, 1994.*

In parallel with the scientific and regulatory processes, the Department of Energy is working to characterize the waste that would qualify for disposal at the WIPP. If the WIPP is approved for permanent disposal, most of the transuranic waste now in storage would eventually be emplaced there. However, there is also a large amount of transuranic material that in its current form contains too much plutonium to be acceptable at the WIPP.

Low-Level Radioactive Waste

As defined by law, "low-level waste" is a catch-all term for radioactive waste that is not high-level waste, transuranic waste, spent fuel, or mill tailings. The Department's policy also allows certain other materials to be managed as low-level waste: small volumes of material used for nuclear research and development, material contaminated with small concentrations (less than 1 ten-millionth of a curie per gram of waste) of transuranics, and small concentrations of naturally occurring radioactive material as well as waste produced in research projects. Virtually any activity involving radioactive materials generates some low-level waste. This waste can include a wide variety of forms and radioactivity levels. The physical forms of low-level waste include rags, protective clothing, contaminated equipment, waste resulting from decontamination and decommissioning, construction debris, filters, and scrap metal.

Low-level waste is also generated by commercial power reactors and facilities producing fuel for them. In addition, it also comes from industrial sources and research laboratories. Another source is the world of medicine, where radioactive isotopes are used for diagnosis and treatment.

Most of the Department's low-level waste has been packaged in drums or boxes and buried in shallow pits and trenches. Approximately 3 million cubic meters has been disposed of in this way.

> ***"Low-level waste" is a catchall term for radioactive waste that is not high-level waste, transuranic waste, spent fuel, or mill tailings.***

This engineered trench for low-level waste contains approximately 30,000 stacked carbon-steel boxes of waste, each box being 4 by 4 by 6 feet. It stopped receiving waste in 1995. In 1996, the site will be backfilled with dirt to form a mound, which will be seeded with grasses and sloped for runoff. Once this trench is closed, the trench-burial of low-level waste here will stop.
Engineered Low Level Trench 4, Solid Waste Management Burial Grounds, Savannah River Site, South Carolina. January 7, 1994.

Newly Generated Low-Level Waste

Although weapons production has been suspended, some low-level waste is still being generated. In fact, low-level waste accounts for more than 80 percent of the Department's newly generated waste, which consists of clothing, tools, and equipment used in cleanup operations, contaminated soils, dismantled buildings and machinery.

To improve efficiency, the Department is stressing waste minimization and early characterization and segregation of waste to reduce the generation of low-level waste requiring disposal. In addition, treatment methods are being improved to reduce waste volumes and provide more stable waste forms. Minimizing waste volume reduces the cost of disposal and extends the life of disposal facilities. More stable waste forms enhance the overall safety of disposal.

Some low-level liquid waste is a byproduct of efforts to consolidate and stabilize high-level waste for permanent disposal. This liquid waste is being stored temporarily, and some of it is being made into a material called "saltstone" (the waste, which contains various metallic salts, is mixed with concrete). This material is being disposed of in vaults designed to isolate it from the environment.

Managing Low-Level Waste

Low-level waste is currently disposed of at the Nevada Test Site, Hanford, the Savannah River Site, Oak Ridge, Los Alamos, and the Idaho National Engineering Laboratory. Unique wastes, including hull sections of decommissioned nuclear submarines, have been shipped to Hanford and the Nevada Test Site for disposal.

In all cases, newer buried low-level waste is required to meet much more stringent disposal standards. In some cases, many former disposal sites are being re-evaluated to decide whether there is economic or environmental justification for digging up and treating the wastes and contaminated soil. For instance, at a trench in Idaho that contained about 38 kilograms of plutonium, the low-level waste was excavated, packaged, and stored for later disposal.

This new vault for storing low-level waste contains 12 large cells, each of them 55 feet long, 150 feet wide, and 30 feet high. The first facility of its kind in the country, this vault system will replace shallow burial in engineered trenches at the Savannah River Site. This vault began storing waste in September 1994. Once full, it will be covered with clay to form a mound with a plant cover. *E Area Vault, Solid Waste Management Division, Savannah River Site, South Carolina. January 7, 1994.*

At the Rocky Flats Plant, about 700,000 gallons of contaminated sludge from five solar evaporation ponds was first consolidated into a single pond and is now being transferred to about 70 large double-walled polyethylene tanks. This program will isolate the material and alleviate concerns that ground and surface waters will be further contaminated while a cost-effective long-term treatment is selected.

Researchers at Rocky Flats are exploring methods of mixing radioactive waste with recycled polyethylene so that it can be poured into drums or other forms for disposal. Large quantities of polyethylene beverage containers are already discarded in landfills, where the longevity of the plastic slows organic decomposition. A combination of low-level waste and waste plastic would take advantage of polyethylene's durability, while reducing waste-plastic in conventional trash landfills.

At Fernald, a vitrification technology is being developed for treating wastes contaminated with uranium and other natural radioactive isotopes. Using a process similar to high-level-waste vitrification, the Department will make wastes into glass pebbles, or "gems," that are much more resistant to leaching than the original waste.

Since the 1980s, the Department has safely operated some below grade containment wells and above grade disposal facilities for low-level solid waste at Oak Ridge. The construction of two new facilities will begin in 1995 and 1998.

The Savannah River Site has constructed and begun operating some low-level-waste vaults to replace the traditional shallow-land-burial trenches.

Some types of low-level waste, such as high-activity waste, require greater confinement than that provided by shallow land burial. To determine a disposal method for these wastes, the Hanford Site and the Nevada Test Site are evaluating the design and use of engineered facilities.

The Z-Area vault for low-level-waste in saltstone form is a 25-foot-tall reinforced-concrete structure 600 feet long and 200 feet wide, housing 12 concrete cells that will be filled with solid grout. The grout is made of a low-level radioactive solution mixed with cement, fly ash, and slag. The chief radionuclides locked into the grout are technetium 99, strontium 90, and cesium 137. Once all 12 cells are filled, the vault will be covered with earth and capped with clay. *Savannah River Site, South Carolina.*

Byproducts of the Cold War

Hull sections of decommissioned nuclear-powered submarines are put in disposal trenches. The used nuclear fuel is removed from the sections of submarine hulls that contain nuclear reactors. The radioactively contaminated hull sections with the defueled reactors inside are then transported by barge to Hanford, where they are placed in a trench for burial. *Trench 94, Hanford Site, Washington. July 12, 1994.*

Use of the thick submarine hull as a disposal container provides extra isolation between the environment and the low-level waste and toxic lead that remain after the reactor fuel is removed. *Trench 94, Hanford Site, Washington. December 20, 1993.*

A burn cage is used at the Pantex site to dispose of items associated with the shipping and handling of high explosives used to make nuclear warheads. Wooden and cardboard crates and other materials contaminated with high explosives are burned inside the cage. *Burning ground, Pantex Plant, Texas. November 18, 1993.*

Hazardous Waste

Although radioactive waste certainly presents hazards, a waste is not legally considered "hazardous" unless it contains other chemicals or exhibits particular characteristics, such as being ignitable or corrosive under some circumstances. This legal distinction is important because if a waste is determined to be "hazardous" under the solid- and hazardous-waste law known as the Resource Conservation and Recovery Act, a rigorous set of regulations applies. Some States, such as Washington, have established additional requirements for other wastes considered "dangerous." A landmark legal case in 1984 determined that hazardous-waste requirements do apply to waste that contains radioactivity as well as hazardous constituents–so-called "mixed waste." The Energy Department has successfully negotiated aggreements with appropriate states to treat these wastes and is committed to complying with these requirements.

The Department's hazardous (non-radioactive) wastes are essentially the same as industrial chemical wastes produced by private corporations and, in much smaller quantities, by most households. They include organic solvents remaining from an incomplete chemical reaction, sludges from degreasing operations, heavy metals from unrecycled batteries. Generally, the Department uses private vendors to remove hazardous waste from its sites and to treat it and dispose of it in compliance with regulations.

Although hazardous waste may present more conventional and familiar risks than the radioactive wastes generated by the Department, it is important to note that safe handling requires substantial expertise and training, and constant vigilance. In the past, like many private companies, the Department has often failed to take adequate precautions in handling, storing, treating, or disposing of hazardous waste. The result is significant environmental contamination that now requires expensive remediation. In some cases, stored waste is discovered for which no records are available to characterize it. These "unknowns" can be among the riskiest wastes to manage.

Like private industry, the Energy Department has learned that the best way to manage hazardous waste is to minimize the amount generated or to eliminate its generation in the first place.

Mixed Hazardous and Radioactive Waste

All high-level waste and most transuranic waste is mixed waste, usually because of the presence of organic solvents or heavy metals in addition to radioactive components. In this discussion, however, the term "mixed waste" is used to mean *low-level* radioactive mixed waste.

The hazardous component of mixed waste is regulated under the Resource Conservation and Recovery Act. In 1992, President Bush amended this act by signing into law the Federal Facility Compliance Act (FFCA), which, among other provisions, expanded the regulation of the Department of Energy's mixed waste. The FFCA made Federal facilities subject to the same fines and penalties as any private corporation if they violate the law. The law also requires the Department to develop plans for mixed-waste treatment, subject to approval of the states or the Environmental Protection Agency.

While the Department increased its compliance with environmental requirements for purely chemically hazardous wastes during the 1970s and 1980s, it accumulated large amounts of mixed waste in storage because of a lack of treatment and disposal facilities. As of 1984, however, the Resource Conservation and Recovery Act required that much of this waste be stabilized in preparation for disposal and not indefinitely stored. The Department is now faced with an enormous challenge–where and how to treat the large backlog of waste.

It may take many years to develop suitable treatment technologies, build facilities, and treat the existing backlog of mixed waste.

A Fernald worker overpacks rusting 55-gallon drums of low-level mixed waste by sealing them inside larger new 85-gallon drums. Some 50,000 deteriorating drums of Fernald waste stored outdoors for many years are being overpacked in a project that began in the late 1980s. *Plant 5, formerly the Metals Production Plant, Fernald Environmental Management Project, Fernald, Ohio. December 28, 1993.*

This incinerator in Oak Ridge burns radioactive and mixed hazardous radioactive wastes. Licensed for operation by the Environmental Protection Agency, it is the only one of its kind in the United States. *Toxic Substances Control Act Incinerator, Oak Ridge, Tennessee. January 10, 1994.*

To develop treatment plans, the Department, in conjunction with the National Governors' Association, has been working closely with the 22 states in which its mixed wastes are stored. New or improved cost-effective technologies also are being pursued. In general, radioactivity was not considered when technologies for commercial hazardous wastes were being developed; however, some can be adapted to deal with it. The Department has used an incinerator at Oak Ridge, Tennessee, to treat some mixed waste, but its technologies are not large or versatile enough for all treatment needs. Alternative, innovative technologies like plasma furnaces, vitrification, and polyethelyene encapsulation, promise to improve performance, reduce risks, and increase economic efficiency beyond the existing technologies of incineration and cementation. However, it may take many years to develop suitable treatment technologies, build facilities, and treat the existing backlog of mixed waste. During that time the Department will work with regulators, Native American Tribes, and the public to develop adequate disposal facilities.

Byproducts of the Cold War

The encapsulation of low-level mixed waste in polyethylene is an innovative waste-handling technology in a pilot phase at Rocky Flats. A heated stream of recycled polyethylene is combined with simulated low-level mixed radioactive waste, encapsulating each particle of waste as the mix is poured into molds. *Building 881, Rocky Flats Plant, Colorado. March 21, 1994.*

The vitrification of low-level mixed waste was demonstrated during the Minimum Additive Waste Stabilization pilot project. This demonstration used nonradioactive waste simulating soils and sludges contaminated with uranium and thorium. It produced several thousand kilograms of thumbnail-sized glass pebbles. This innovative technology makes wastes more stable while reducing waste volume and rendering them safer for disposal. *MAWS Facility, Fernald Environmental Management Project, Fernald, Ohio. December 28, 1993.*

Other Materials in Inventory

With the end of the Cold War, many valuable materials once used as primary materials or recycled back into the production cycle are no longer needed for their original purposes. Such materials range from plutonium residues in gloveboxes to large cylinders of depleted uranium gas to huge piles of contaminated and uncontaminated scrap metals. Some of these materials can be recycled or reused; others may no longer have an economically feasible use. In any case, the Department is working to ensure these materials are managed safely and in an environmentally sound manner.

The Department owns thousands of 10- and 14-ton-capacity steel cylinders filled with depleted uranium hexafluoride from uranium enrichment. During the Cold War, some of this depleted uranium was used to make nuclear weapons parts, targets for plutonium reactors, "tank killer" bullets, and armor-plating used in the 1991 Gulf War. The Department is now working with state regulators and other interested parties to determine the best options for managing its remaining inventory of depleted uranium hexafluoride.

Many thousands of tons of scrap steel, copper, nickel, and other metals are located at sites throughout the nuclear weapons complex. Some of this scrap metal is radioactively contaminated. The Department's policy is to assume that scrap metal is contaminated unless proven otherwise. The Department is investigating ways to recycle some of these materials.

The Department also owns a variety of hazardous chemicals throughout the complex–from small vials containing toluene at the Los Alamos and Livermore Laboratories to large tanks of radioactively contaminated nitric acid at the Hanford and the Savannah River Sites. Many chemicals and chemical residues were left in containers or in process lines when the production of nuclear weapons came to a halt. The strategy for managing these chemicals emphasizes (1) the removal of excess or unneeded chemicals, (2) proper storage, and (3) improved inventory tracking and control.

The inventory includes a variety of other materials like lead, concrete shielding, lithium, and sodium. The Department must ensure that all of these materials are managed safely; it intends to work with regulators and other citizens to determine long-term options for these materials.

This yard for contaminated scrap metal contains heaps of slightly radioactive scrap steel, ferrous metal, and nickel-plated metal left over from upgrades and renovations to the K-25 Gaseous Diffusion Plant at Oak Ridge over the years. *K-25 Scrapyard, Oak Ridge, Tennessee. January 10, 1994.*

Byproducts of the Cold War

A crate of mercury flasks, which were used for lithium-enrichment operations at the Y-12 plant. Lithium must be enriched before it can be used as a target inside a reactor to produce tritium for nuclear weapons. Lithium enrichment was shut down in 1962, leaving about 35,000 areas where mercury remained in the operational equipment; some of it has migrated into the environment. *Y-12 Plant, Oak Ridge, Tennessee. January 11, 1994.*

These Mark 31 depleted uranium target element inner cores are part of the Department's large inventory of nuclear materials left "in the pipeline" when the Cold War ended. These materials are no longer needed for their originally intended use. The Department will work with regulators and other interested parties to determine how materials like these should be managed. *Plant 6, formerly the Metals Fabrication Plant, Fernald Environmental Management Project, Fernald, Ohio. December 28, 1993.*

Aerial view of the F Area at the Savannah River Site. This is one of the two chemical separation areas at the site. It covers just over half a square mile. All facilities shown on the facing page are within 5 miles of this spot. Visible in this photograph are the F Area Seepage Basin before closure (lower right), the F Area Tank Farm (group of circles near the bottom center), the 242-F Evaporator (between the two rows of tank tops), and the 221-F Canyon (long building above parking lot). *F Area, Savannah River Site, South Carolina. August 6, 1983.*

Waste-Handling Complications

It was necessary to construct a vast network of industrial facilities to mass produce materials and parts for nuclear weapons. Similarly, another chain of plants and processes is needed to contain, stabilize, treat, store, and prepare for disposal the resulting radioactive wastes. Each process leads to others, and each generates waste that must be handled. The cleanup of contamination also generates wastes that must be managed carefully, and each step in this process typically generates more waste.

There are thousands of industrial buildings and structures throughout the Department's sites. To the uninitiated observer, there is no apparent relationship among them. Yet each is inextricably linked to the others. An understanding of these connections is critical to the success of the environmental management mission. An example of how these connections interact, how one process or facility leads to another, is perhaps best illustrated by the chain of processes at the Savannah River Site, seen on page 59.

An understanding of the connections between facilities is critical to the success of the Environmental Management mission.

Byproducts of the Cold War

12. To store the waste from the Saltstone Facility, we built the grout vaults.

1. To produce plutonium for nuclear warheads, we placed uranium "targets" in nuclear reactors and bombarded them with neutrons.

2. To extract plutonium from reactor targets, we built reprocessing "canyons," which generated liquid high-level radioactive waste.

11. To solidify the volume of liquid waste not processed by the Defense Waste Processing Facility, we built the Saltstone Facility.

3. To store the liquid high-level radioactive waste from the reprocessing canyons, we built underground storage tanks.

Savannah River Site Connections

10. To reduce the volume of liquid high-level waste, we built the In-Tank Precipitation Facility.

4. To handle the low-level waste from reprocessing, we built burial grounds.

9. To stabilize the liquid high-level radioactive waste in the storage tanks, we built the Defense Waste Processing Facility.

5. To make space for more liquid high-level radioactive waste in the storage tanks, we built evaporators to reduce the waste volume.

8. To provide an alternative for discharging effluent from the evaporators, we built the Effluent Treatment Facility.

7. To arrest spreading ground-water contamination from waste poured into the seepage basins, we built clay caps over them and installed pumping wells.

6. To dispose of wastewater that the evaporators removed from the high-level waste tanks, we built seepage basins.

Closing the Circle on the Splitting of the Atom

IV. Contamination and Cleanup

Every site in the complex is contaminated to some extent with radioactive or other hazardous materials. This contamination occurs not only in buildings; it is also found in soil, air, ground water, and surface water at the sites. Some sites and many of the buildings that were used during the Manhattan Project have already been cleaned up. However, most sites have significant and complicated problems that have been compounded over several decades.

For example, at the Hanford Site in the State of Washington, tritium has been detected in ground water, and high-level waste has leaked from storage tanks. At Oak Ridge, Tennessee, an estimated 1,000 tons of mercury have been released into the environment. At Fernald, Ohio, several hundred tons of uranium dust were emitted into the atmosphere, and drinking-water wells were contaminated with uranium. Traces of plutonium have been found in the soil and sediments around the Rocky Flats site in Colorado.

Fallout from aboveground nuclear tests in the United States and other countries has radioactively contaminated the atmosphere surrounding the entire Earth. Contamination with radioactive iodine released from early operations at the Hanford Site in Washington was also widespread. The large buildings used for reprocessing spent fuel at the Hanford Site and the Savannah River Plant in South Carolina are so contaminated with radioactive materials that decontamination must be done by remote control to protect the workers.

Decontamination worker at Hanford's UO_3 Plant scrapes down a workshop interior to remove low-level radioactive contamination on floor surfaces. *UO_3 Plant, Hanford Site, Washington. July 11, 1994.*

Every site in the complex is contaminated to some extent with radioactive or other hazardous materials.

Workers are decontaminating equipment used to move contaminated soil at the Weldon Spring site. Facilities at this site once performed many of the same functions as the Fernald Plant. *Weldon Spring site, Missouri. January 29, 1994.*

Actions in Cleanup

To understand environmental remediation, it is useful to look at the sequence of actions that are undertaken at contaminated sites:

1. A site is "characterized" by collecting data from soil and sampling wells, for example, in order to understand the nature and extent of contamination, its potential consequences, and the response alternatives. Computer modeling is often used to help estimate the spread of contamination.

2. The spread of contaminants is contained by using proven methods to slow or stop it.

3. Buildings are decommissioned and decontaminated. The first priority is safely maintaining the buildings before final disposition. When resources are available, the buildings are cleaned and then, in most cases, demolished.

4. The site and land are cleaned up by removing, consolidating, and stabilizing contaminants; the site is then prepared for future use.

In daily practice, contamination is addressed first through prevention, including the sound management of waste and other contaminants. When contamination does occur, cleanup options must be evaluated to avoid actions that might compound the problem. Finally, decontamination is undertaken where practical.

Deciding When and How To Take Action

The Department of Energy is committed to "moving dirt more than paper" and making progress. It is also committed to investing in technology that leads to more effective and efficient treatment. Although aggressive action sounds appealing, cleanup and decontamination are not so simple.

For example, while cleaning up contaminated soil, water, or buildings, workers will likely generate huge amounts of new waste that will require adequate storage, treatment, and disposal.

Another problem is that, by their very nature, radioactive materials and heavy metals cannot be

Contamination and Cleanup

destroyed. Over time–from fractions of a second to tens of thousands of years–radioactivity decays naturally. Meanwhile, radioactive wastes must be contained, stabilized, or moved to a safer place.

If contamination is not removed or stabilized, workers or the public could be exposed to radiation and other hazards. In some cases greater hazards can result from cleanup. One of the largest offsite releases of plutonium from the Rocky Flats Plant stemmed from an effort to scrape up contaminated soil on a hillside where drums filled with plutonium-contaminated waste had leaked. While the area was being scraped, strong winds carried plutonium-contaminated dust across a large area of nearby land. Cleanup workers were especially at risk.

Finally, some sites appear too severely or broadly contaminated to be cleaned up by the methods, resources, and funds currently available. Although technology development might help, no cost-effective remedies are on the horizon. Moreover, at many sites the benefits of cleanup are not worth the additional damage that might be inflicted on the environment or the potential risks to cleanup workers.

The Department has made significant progress. Many Manhattan Project facilities and 5,000 vicinity properties have already been cleaned up.

Specific sites that fit these categories cannot be easily listed, but they clearly exist. For example, hundreds of nuclear detonations left residual radioactivity at the Nevada Test Site. Most of this radioactivity is in highly inaccessible underground locations. There is no cost-effective technology for decontaminating such sites. Other facilities face similar difficulties. Many such sites will be isolated and monitored until practical cleanup methods are developed or until risks from the contaminants have diminished to a point where the land can be used again.

The White Oak Creek embayment is sited where the Clinch River meets White Oak Creek, whose waters flow through the site of the Oak Ridge National Laboratory. When creek waters leave the site, they are contaminated with cesium 137, strontium 90, and PCBs. Until 1991 there was a cable with a warning sign at this point. In 1992 the Department constructed a state-of-the-art sediment-retention dam that uses interlocking sheets of metal driven into bedrock to retard the flow of water so that contaminated sediments can settle behind the dam. *Oak Ridge, Tennessee. January 11, 1994.*

It is also true that while cleaning up some parts of sites will benefit ecosystems, other remediation efforts might damage them. At the Savannah River Site, a 2,600-acre lake used for cooling a production reactor became contaminated, primarily with cesium 137, a highly radioactive isotope. One remedy would be to drain the lake, then scrape up and contain the contaminated sediments. However, that action would destroy a valuable habitat for migratory birds and other animals. It would also expose workers and the public to greater risks. A better approach in this case might be to fence off an area around the lake for 100 to 200 years, allowing the sediment's radioactivity to decline by 10 to 100 times.

Progress in Cleanup

The Department has made significant progress in cleaning up sites and facilities. Many of the sites involved in the early stages of the Manhattan Project have been cleaned up and their buildings have been decontaminated or demolished under the Formerly Utilized Sites Remedial Action Program. Although most of these facilities are relatively small, some had been heavily contaminated. Cleanup has been completed at 21 such formerly used plants in Illinois, New York, New Jersey, and elsewhere.

Other contaminated sites have demanded an immediate response because people live in or on them, or because large concentrations of hazardous material were exposed to the elements.

For example, uranium-mill tailings emit radon gas, an identified health hazard. Large volumes of sandy radioactive tailings were left in open piles, subject to rain and wind, and some of this material was used for constructing roads, houses, schools, and other buildings. About 5,000 of these vicinity properties have been cleaned up under the Uranium Mill Tailings Remedial Action Program. This program has made steady progress in consolidating and capping huge tailings piles at dozens of former mill sites in several western states. Sixteen of the 24 mill sites have been remediated to date.

At the Y-12 site in Oak Ridge, Tennessee, several large settling ponds were part of a waste-water-treatment facility for acids and organic wastes containing uranium. Beginning in 1985 the liquid in these ponds was treated to remove contaminants and the ponds were drained and

During the cleanup of mercury contamination, this worker uses a special suit and respirator for protection against mercury-vapor poisoning. Many tons of mercury were released to the environment at Oak Ridge's Y-12 Plant during lithium-enrichment operations. Enriched-lithium targets are needed to make tritium, a radioactive gas used in nuclear weapons. *Oak Ridge, Tennessee. January 11, 1994.*

capped. Since 1990 the area has been safely used as a parking lot (see page 69).

A hillside, called the 881 Hillside, within the site boundaries of the Rocky Flats Plant in Colorado, was contaminated with a variety of radioactive isotopes, toxic metals, solvents, and petroleum products. The Department of Energy installed monitoring wells that identified the potential for releasing contaminants into offsite ground water and surface streams. Along the downhill edge of the site, an impermeable barrier and a "french drain" collection system were installed. Contaminated ground water has been pumped out of the collection system and treated. Cleanup workers also removed "hot spots" of radioactively contaminated soil and stored it in drums.

Challenges To Be Met

The Department faces more-expensive, longer-term decontamination challenges than the examples given above. Decontamination is needed at several thousand facilities that have been declared surplus. These include more than a dozen large reactor buildings, nine chemical separation plants, three vast uranium-enrichment complexes, and an array of smaller plants. The interiors of some of these buildings are too radioactive for unshielded workers to enter them. Robotics technology once used for production is now being adapted for decontamination and dismantlement work in these plants.

Cleanup planning goes hand-in-hand with facility transition and maintenance. To prevent accidental releases of radioactive materials, and to minimize hazards to cleanup workers, it is important to keep these buildings in stable condition as cost effectively as possible.

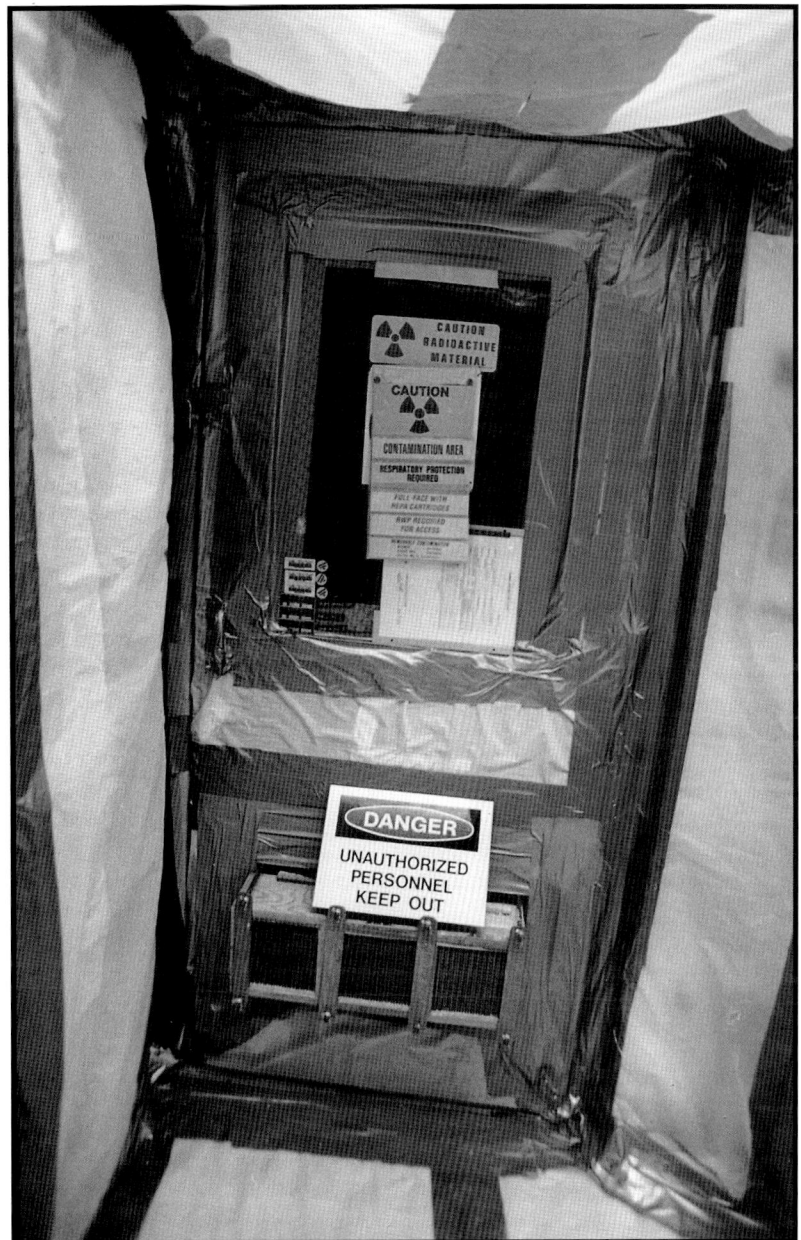

The sealed door to an "infinity room" at Rocky Flats. More than 20 such rooms have been contaminated by releases during plutonium operations at the site. The rooms are called "infinity rooms" because the rates of alpha radiation are too high for standard monitoring equipment to measure. The radioactivity in these rooms is nearly 25,000 times natural background. *Building 776/777, Rocky Flats Plant, Colorado. March 18, 1994.*

Although aggressive action sounds appealing, cleanup and decontamination are not so simple.

Ventilation ducts contaminated with plutonium in dust, oxides, and smoke exhausted from gloveboxes in the pyrochemistry area of Rocky Flats. When a buildup of plutonium becomes too great, it can pose a criticality threat. The buildup in these ducts was close to the limit for such a threat. *Building 776, Rocky Flats Plant, Colorado. December 20, 1993.*

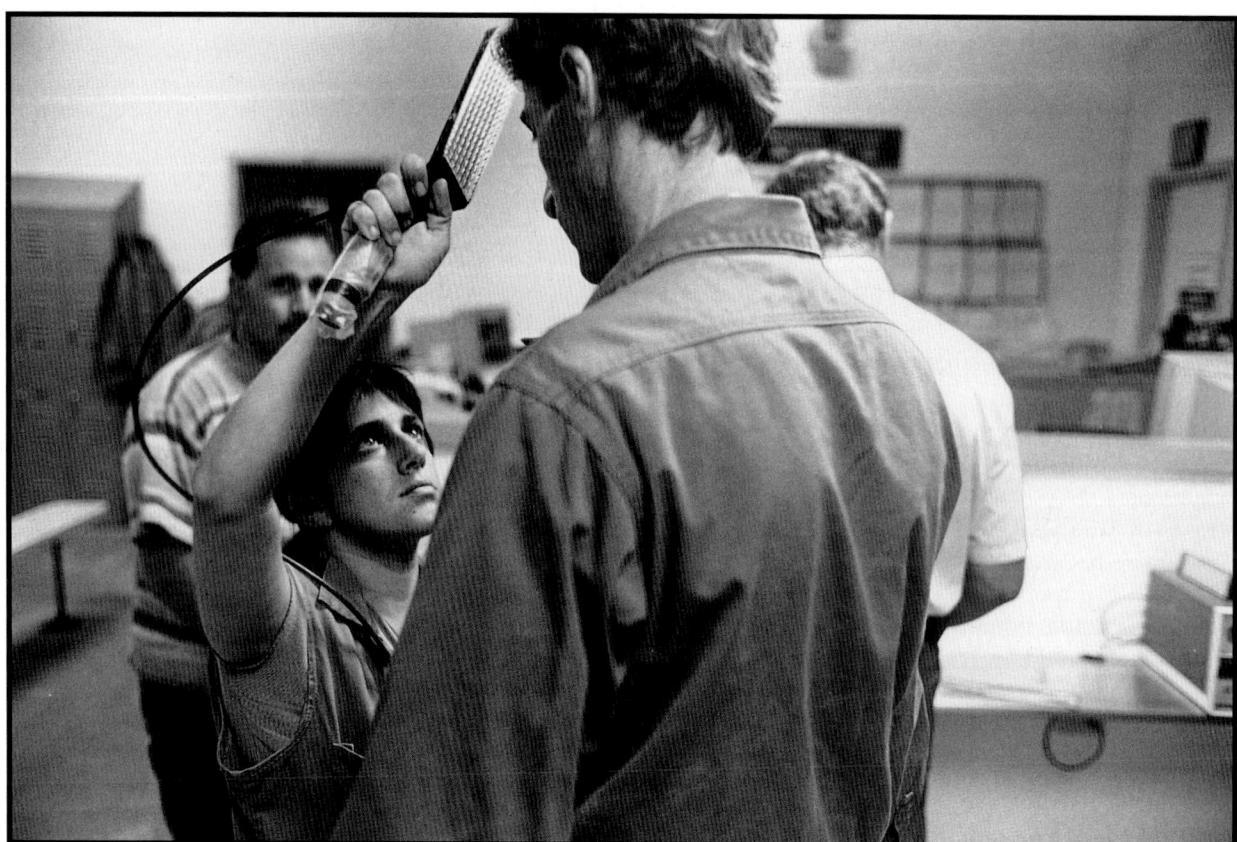

Health-physics technician conducts a whole-body survey for potential radioactive contamination. She slowly moves a detection instrument over a worker, holding the meter within a quarter of an inch of his body. *Plutonium Finishing Plant, Hanford Site, Washington. December 20, 1993.*

Improving Performance

Along with radioactive isotopes, toxic metals and organic chemicals can also be difficult to remove from facilities, soils, and ground water. Some large buildings at Oak Ridge, Tennessee, became heavily contaminated with mercury during lithium enrichment operations. The leftover mercury used in this process is being gradually accounted for and stabilized. Because high concentrations of mercury are very toxic, workers in the area must wear special clothing and respirators, and they must proceed cautiously. The environmental management program at Oak Ridge is mapping this contamination and taking steps to prevent its further spread.

As workers and contractors become more proficient at environmental restoration, they are finding creative ways to improve performance. A good example is Hanford's T Plant, which was a reprocessing plant that extracted the plutonium used for the Trinity test, the Nagasaki bomb, and other early weapons. This huge building is now being used for cleaning equipment with high-activity contamination. Using an already contaminated building for such a purpose avoids the costly construction and decontamination costs of a new facility.

The Department is investing in technologies to make cleanup more effective. In this new era of openness and public involvement, citizens and the government can work together to ensure that progress continues and that environmental and public-health risks are reduced and workers are protected.

As workers and contractors become more proficient at environmental restoration, they are finding creative ways to improve performance.

Wastes from the earliest days of the Manhattan Project were buried in a 21.7-acre field just north of the St. Louis Airport, starting in 1946 and continuing for about 10 years. Today water draining from the field into a ditch bordering the site gives radiation readings 10 to 15 times higher than the natural background. Certain contaminants, such as thorium 230, tend to cling to the sediments in these ditches and have accumulated to significantly greater concentrations than in the water. *St. Louis Airport FUSRAP Site, Missouri. January 30, 1994.*

BEFORE – The Shippingport atomic power station before decomissioning. Built in 1957, Shippingport was the first large-scale nuclear power plant in the world. *Shippingport, Pennsylvania.*

AFTER – The Shippingport site after decontamination and decommissioning by the Department of Energy in 1990. This was the first complete decontamination and decommissioning of a power-producing reactor in the nation. *Shippingport, Pennsylvania.*

Contamination and Cleanup

BEFORE – These four ponds received wastewater until 1985 from operations at the Y-12 Plant. *Oak Ridge, Tennessee.*

AFTER – A parking lot is now located where the four ponds shown above once stood. The liquids in these ponds were treated to remove contaminants beginning in 1985; the ponds were then drained and capped with asphalt. The project was completed in 1990. *Oak Ridge, Tennessee.*

The Department owns more than 2,000 contaminated facilities that will require decontamination and decommissioning.

Robotics technology once used for production is now being adapted to clean up contaminated facilities.

Moving Forward

The Department of Energy is decontaminating and demolishing old buildings, pumping and treating contaminated ground water, packaging contaminated soils, capping old dumping grounds to keep rainwater out, and moving drums of waste into sheltered structures. Many of these activities do not provide permanent solutions. Often they are the least costly and least risky means of holding contamination in place while priorities are set and decisions are made for the long term.

Affected citizens and workers, the Congress, Native American Tribes, and State and Federal regulatory agencies are actively participating in these decisions. They are addressing some of the following difficult questions:

How clean is clean? Given that radiation is everywhere, how do we decide when additional manmade radiation is a problem and when it is not? There is no universal right answer. This decision depends on site characteristics, the costs of remediation, and the use of the land. However, many immediate hazards are recognized, and the Department of Energy is addressing urgent risks on the basis of what is known rather than waiting for more information at the risk of increasing potential adverse impacts.

Should we decontaminate sites now or wait for better technology? The Department of Energy is working to evaluate emerging cleanup methods. It supports reseach and development in cases where both risks and current remediation costs are high, and it is developing contract incentives to encourage innovation and efficiency. However, some of the best technologies currently available preclude further treatment in the future.

How much scientific certainty is needed? Risk assessment is subject to many unknowns. How much additional research is needed to reduce uncertainty? How do we decide what to do with imprecise data? When do we stop studying and start acting?

What are the benefits of cleanup? While the financial cost of responsible environmental management can be calculated, its benefits are difficult to put in dollar terms. The positive results of cleanup can include reductions in worker and public risk as well as the value of land and facilities turned over to public or private use.

Contamination and Cleanup

Demolition of a 456-foot-long structure built in 1943 brings an end to one of the original buildings at the Hanford Site. The building housed 1.7-million-gallon water-storage tanks that fed the cooling pumps of the Hanford B Reactor. Decommissioning crews removed the tanks, knocked down concrete walls, took out underground piping, filled in piping tunnels, and then collapsed the steel structure with explosive charges. Demolishing this building reduced hazards as well as surveillance and maintenance costs. Noncontaminated concrete and steel are recycled. *Hanford Site, Washington. December 1993.*

Workers remediate the 881 Hillside at Rocky Flats, an area that became heavily contaminated with toxic and radioactive substances. As part of the remediation action at the site, workers cleaned up six "hot spots" of highly radioactive contaminated soil. *Rocky Flats Plant, Colorado. September 1994.*

This exhaust stack was used to control emissions from the Plant 9 facility, where enriched uranium materials were processed. The malfunctioning of systems like this resulted in several releases of uranium dust, totalling several hundred tons, to the environment outside the plant buildings over the course of operations. *Fernald Plant, Ohio. December 30, 1993.*

Lisa Crawford: A Citizen of Fernald, Ohio

Lisa Crawford's husband Ken worked at a General Motors plant in rural Ohio outside Cincinnati. The Crawfords, with their two-year-old son Kenny, moved to Fernald in 1979. They rented a farmhouse across the road from a plant with red-and-white checkered water towers called the "Feed Materials Production Center." "Like a lot of people around here," Lisa said, "we thought it made cattle feed or dog chow."

In late 1984, a local journalist reported that the plant had released a large amount of radioactive dust into the air and that three local wells were contaminated with uranium. One of the wells served the Crawford farmhouse. Lisa and her husband learned that the Feed Materials Production Center made components for nuclear weapons. They also found out that the Department of Energy had been aware that their well was contaminated as early as 1981—yet sent annual reports to their landlord saying tests had proved the water safe.

Soon after discovering that her family had been using contaminated well water for years, Lisa helped found a community organization called Fernald Residents for Environmental Safety and Health, or FRESH. In January 1985 she and her husband filed a $300 million class-action suit on behalf of the 14,000 citizens living within 5 miles of the plant against the contractor for the Department of Energy site, National Lead of Ohio.

Three years after the lawsuit was filed, the Department of Energy acknowledged that there had been uranium leakage at the plant since it had opened in 1951. In all, more than 100 tons of uranium dust had been released into the air, and more than 70 tons had been dumped into a local river. The ground water was found to be contaminated with chlorides, nitrates, fluorides, and uranium. In 1989, the lawsuit was settled, and the Department paid $78 million in damages to the citizens of Fernald.

In the late 1980s, the Fernald site shut down its weapons-production operations completely, and a new contractor took over the site. The Department of Energy has begun to clean up the site, a task expected to take several years.

Lisa Crawford and FRESH have been instrumental in shaping public involvement at Fernald. Working with site personnel, thay have found innovative ways to achieve meaningful public participation. "Once trust is taken away," Crawford said, "it's very hard to get it back. DOE must continue to work cooperatively with the community and clean up the Fernald site. Then, and only then, will the possibility of trust be restored."

Contamination and Cleanup

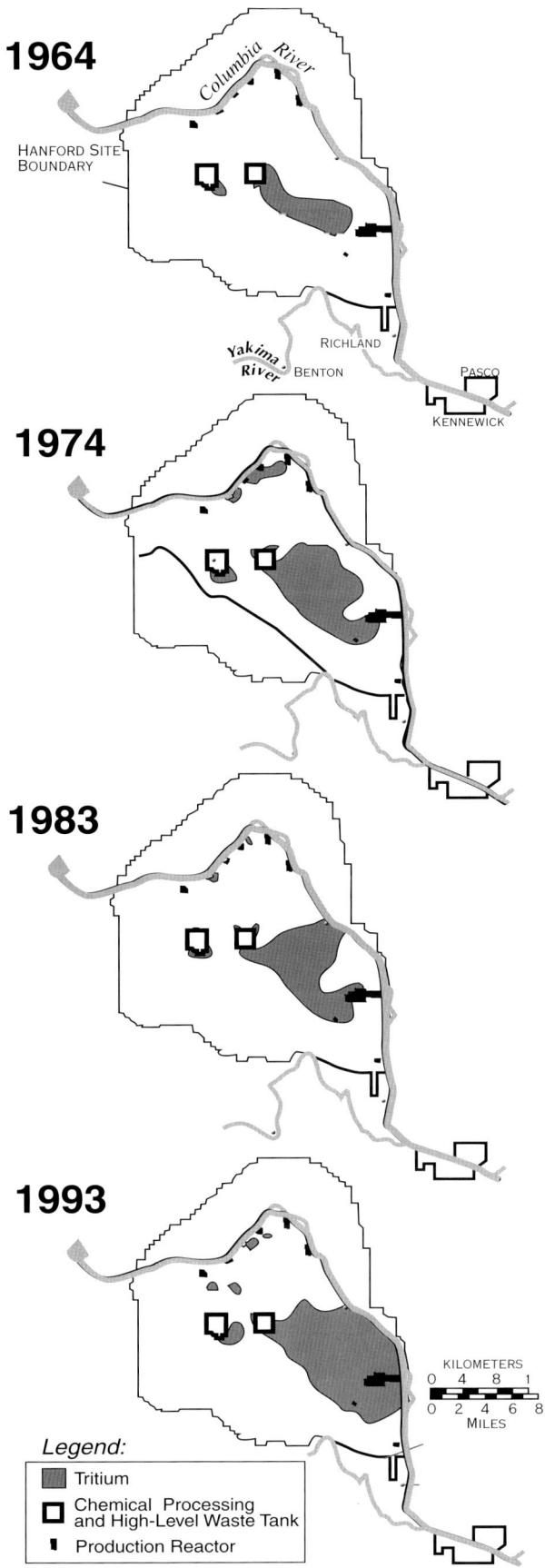

Spreading tritium contamination at the Hanford Site in Washington. The shaded areas on these maps show how tritium contamination in concentrations above safe drinking-water standards has spread over time.

Dose Reconstruction: Estimating Past Human Exposures

Releases of radioactive materials associated with nuclear weapons production at sites throughout the weapons complex have aroused concern about potential public-health consequences. No one knows exactly who among the general public was exposed to how much radioactivity during the Cold War or what actual health impacts resulted. The Centers for Disease Control has undertaken "dose reconstruction" studies around several major Department of Energy facilities to gain a clearer understanding of potential health effects through epidemiological research. Efforts begin with trying to determine how much radiation was received by citizens living near nuclear weapons sites.

One of the earliest and most extensive research efforts began at Hanford in Washington in 1986. After the DOE assembled hundreds of documents addressing the environmental impacts of its operations from 1945 to 1985, a committee of representatives from Washington, Oregon, the Yakima Indian Nation, the Confederated Tribes of the Umatilla Indian Reservation, and the Nez Perce Tribe concluded that radioactive releases and biological pathways should be studied in order to "reconstruct" potential radiation doses to the public. The objectives of the Hanford Environmental Dose Reconstruction Project are to estimate the radiation doses that populations could have received from nuclear operations at Hanford since 1944, and to make public all the information used in the project. In order to obtain dose estimates from past radioactive releases, historical data are being identified, reviewed, and analyzed in order to understand atmospheric, river, and ground-water conditions that affected the transport of radioactivity from operating facilities to offsite populations. The types and quantities of radioactive materials emitted by Hanford's operations are also being evaluated. As information on population distributions, agricultural practices, and eating habits is obtained, the migration of radionuclides through environmental pathways to regional populations will be modeled.

To provide independent technical direction to the effort, professors from area universities selected a Technical Steering Panel from a list of candidates. The technical steering panel currently has nine members and includes representatives from a range of organizations. All project reports that have been approved by the technical steering panel and references used in the reports are being placed in a local public reading room.

Dose reconstruction studies at Hanford and other sites will help build the informational foundations for sound risk assessment. The experience gained in these pioneering efforts should be valuable in a wide range of environmental projects.

The U.S. DOE Environmental Management Program: Responsibilities from Coast-to-Coast and Beyond

This remediated railroad spur in Maywood, New Jersey was radioactively contaminated with thorium unloaded at the site and taken to a nearby factory in Wayne. The thorium was used to produce mantles for gas lanterns. *December 10, 1993.*

DEFENSE SITES

Alaska
 Amchitka Island Test Site, Amchitka Island

Arizona
 Monument Valley (uranium mill tailings)
 Tuba City (uranium mill tailings)

California
 Lawrence Livermore National Laboratory, Livermore (Main site and Site 300)
 Oxnard Site, Oxnard
 Salton Sea Test Base, Imperial County
 Sandia National Laboratories, Livermore
 University of California, Gilman Hall, Berkeley

Colorado
 Durango (uranium mill tailings)
 Grand Junction (uranium mill tailings)
 Grand Junction vicinity properties (uranium mill tailings)
 Gunnison (uranium mill tailings)
 Maybell (uranium mill tailings)
 Naturita (uranium mill tailings)
 New Rifle Mill, Rifle (uranium mill tailings)
 Old Rifle Mill, Rifle (uranium mill tailings)
 Old North Continent, Slick Rock (uranium mill tailings)
 Rocky Flats Environmental Technology Site, Golden (formerly Rocky Flats Plant)
 Union Carbide, Slick Rock (uranium mill tailings)

Connecticut
 Combustion Engineering Site, Windsor
 Seymour Specialty Wire Co., Ruffert Building, Seymour

Florida
 Peak Oil Petroleum Refining Plant, Largo
 Pinellas Plant, St. Petersburg
 4.5 acre site, St. Petersburg

Hawaii
 Kauai Test Facility, Kauai

Idaho
 Idaho National Engineering Laboratory, Idaho Falls
 Lowman (uranium mill tailings)

Illinois
 Granite City Steel, 1417 State St., Granite City, (Formerly General Steel Castings Corp.)
 Illinois National Guard Armory, 52nd Street & Cottage Grove Ave., Chicago
 University of Chicago: New Chemistry Laboratory and Annex, West Stands (Stagg Field), Ryerson Physical Laboratory, Eckhart Hall, Kent Chemical Laboratory and Annex, Ricketts Laboratory

Iowa
 Ames Laboratory, Ames

Kentucky
 Maxey Flats, Hillsboro (LLW Disposal Site)
 Paducah Gaseous Diffusion Plant

Maryland
 W.R. Grace & Co., Building No. 23, Curtis Bay

Massachussetts
 Chapman Valve Building 23, Indian Orchard
 Shpack Landfill, Norton and Attleboro
 Ventron Corp., Beverly (formerly Metal Hydrides Corp.)

Michigan
 General Motors, 1450 East Beecher St., Adrian, (Formerly Bridgeport Brass Co.)

Missouri
 Kansas City Plant, Kansas City
 Latty Avenue Properties, 9200 Latty Ave., Hazelwood
 Weldon Spring Site, Weldon Spring
 St. Louis Airport Site, St. Louis
 St. Louis Airport Vicinity Properties, St. Louis
 Mallinckrodt Chemical Co., 65 Destrehan St., St. Louis

Nevada
 Nevada Test Site, Mercury
 Tonopah Test Range, Nellis Air Force Base, Tonopah

New Jersey
 Chambers Dye Works, DuPont & Co., Deepwater
 Kellex/Pierpont site, NJ Route 440 & Kellog St., Jersey City (Kellex Corp.)
 Middlesex Municipal Landfill, Middlesex
 Middlesex Sampling Plant, 239 Mountain Ave, Middlesex
 New Brunswick Laboratory, New Brunswick

New Mexico
 Ambrosia Lake (uranium mill tailings)
 Acid/Pueblo Canyon, Los Alamos
 Bayo Canyon, Los Alamos
 Chupadera Mesa, White Sands Missile Range, (Trinity test fallout)
 Holloman Air Force Base, Albuquerque
 Inhalation Toxicology Research Institute (ITRI), Albuquerque
 Los Alamos National Laboratory, Los Alamos
 Pagano Salvage Yard, Los Lunas
 Sandia National Laboratories, Albuquerque
 Shiprock (uranium mill tailings)
 Waste Isolation Pilot Plant, Carlsbad
 South Valley Site, Albuquerque

New York
 Ashland Oil Co., Tonawanda
 Baker and Williams Warehouses, New York
 Bliss & Laughlin Steel, 110 Hopkins St. Buffalo
 Colonie Interim Storage Site, Central Ave., Colonie
 Separations Process Research Unit, Knolls Atomic Power Laboratory, Schenectady
 Linde Air Products, Tonawanda
 Niagra Falls Storage Site, Lewiston
 Niagra Falls Storage Site Vicinity Properties, Lewiston
 Seaway Industrial Park, Tonawanda

North Dakota
 Belfield (uranium mill tailings)
 Bowman (uranium mill tailings)

Ohio
 Alba Craft, 10-14 West Rose Ave, Oxford
 Associated Aircraft and Tool Manufacturing, 3660 Dixie Highway, Farfield
 B&T Metals, 425 West Town St. Columbus
 Baker Bros., 2551-2555 Harleau Place, Toledo
 Fernald Environmental Management Project, Fernald (formerly Feed Materials Production Center)
 HHM Safe Site, Hamilton
 Luckey Site, 21200 Luckey Rd., Luckey
 Mound Plant, Miamisburg
 Painesville Site, 720 Fairport-Nursery Rd., Painesville
 Portsmouth Gaseous Diffusion Plant
 Reactive Metals, Inc. (RMI), Ashtabula

Oregon
 Albany Research Center, Albany
 Lakeview (uranium mill tailings)

Pennsylvania
 Aliquippa Forge, Aliquippa
 Bettis Atomic Power Laboratory, West Mifflin
 Canonsburg (uranium mill tailings)
 C.H. Schnoor, Springdale

South Carolina
 Savannah River Site, Aiken

South Dakota
 Edgemont Vicinity Properties (uranium mill tailings)

Tennessee
 Elza Gate Site, Melton Dr., Oak Ridge
 Oak Ridge K-25 Site, Oak Ridge
 Oak Ridge National Laboratory, Oak Ridge
 Y-12 Plant, Oak Ridge

Texas
 Falls City, (uranium mill tailings)
 Pantex Plant, Amarillo

Utah
 Green River (uranium mill tailings)
 Mexican Hat (uranium mill tailings)
 Monticello Millsite and Vicinity Properties (uranium mill tailings)
 Salt Lake City (uranium mill tailings)

Washington
 Hanford Site, Richland

Wyoming
 Riverton (uranium mill tailings)
 Spook (uranium mill tailings)

South Pacific Ocean
 Bikini Island
 Enewetak Atoll

NONDEFENSE SITES

Alaska
 Cape Thompson (Project Chariot)

California
 Stanford Linear Accelerator, Stanford
 Energy Technology Engineering Center (ETEC), Santa Susanna
 General Atomics, La Jolla
 General Electric Vallecitos Nuclear Center, Vallecitos
 Laboratory for Energy-Related Health Research (LEHR), Davis
 Lawrence Berkeley Laboratory, Berkeley
 Rockwell International (Formerly Atomics International), Canoga Park
 Santa Susanna Field Laboratory, Santa Susanna,

Colorado
 Project Rio Blanco peaceful nuclear explosion site, Rifle
 Project Rulison peaceful nuclear explosion site, Grand Valley

Idaho
 Argonne National Laboratory-West, Idaho Falls

Illinois
 Argonne National Laboratory-East, Lemont
 Dow Chemical Co., College & Weaver Streets, Madison
 Fermilab, Batavia
 Site A/Plot M, Palos Forest Preserve, Cook County

Kentucky
 Paducah Gaseous Diffusion Plant, Paducah

Mississippi
 Salmon peaceful nuclear explosion site, Hattiesburg

Montana
 Component Development and Integration Site, Butte

Nebraska
 Hallam Nuclear Power Facility, Lincoln

Nevada
 Project Faultless peaceful nuclear explosion site, Central Nevada Test Area Tonopah
 Project Shoal peaceful nuclear explosion site, Fallon

New Jersey
 Maywood Chemical Works, Maywood
 Princeton Plasma Physics Laboratory, Princeton
 Wayne Interim Storage Site, 868 Black Oak Ridge Rd., Wayne

New Mexico
 Project Gnome peaceful nuclear explosion site, Carlsbad
 Project Gasbuggy peaceful nuclear explosion site, Farmington

New York
 Brookhaven National Laboratory, Upton (Long Island)
 West Valley Demonstration Project, West Valley

Ohio
 Battelle Columbus Laboratories, Columbus
 Piqua Nuclear Power Facility

Pennsylvania
 Shippingport Atomic Power Station

Puerto Rico
 Center for Energy & Environmental Research, Mayaguez

V. An International Perspective

Nuclear weapons materials, parts, and production technologies are fundamentally the same worldwide. The wastes that nuclear weapons industries produce are essentially the same as well. At present, five nations are considered the "declared" nuclear weapons states: the United States, Russia, Britain, France, and China.

About 98 percent of nuclear weapons production occurred in the United States and the former Soviet Union, and the quantities of waste and contamination in those countries correspond roughly to the total number of weapons produced. Most of the waste and contamination from nuclear weapons production in these countries resulted from routine operations, rather than from accidents. In Britain, France, and China, waste accumulations are smaller.

Although nuclear weapons were deployed in several Soviet republics, all the major facilities of the former Soviet nuclear weapons complex are in Russia, except for a nuclear test site and a uranium metallurgy plant, both in Kazakhstan. The Russian production plants are similar in number and scale to those of the United States, but the complex in Russia was organized somewhat differently. Furthermore, Russian production reactors were also used to generate electricity and heat for civilian uses and for this reason were not shut down after the arms race ended.

Ironically, the Russian nuclear weapons complex now has less waste in storage than does the United States because large quantities of its high-level waste (as much as 1.7 billion curies) were poured into rivers and lakes or injected deep underground rather than stored in tanks. Widespread waste discharges have left the Russians with much larger areas of contamination than those in the United States.

Tomsk-7 is a Russian plutonium processing facility in Siberia. It is the site of plutonium-production reactors and chemical separation plants, much like the Hanford and Savannah River Sites in the United States. *Near the city of Tomsk, Russia. April 20, 1993.*

Women from the village of Muslyumovo in the southern Ural Mountains stand on the banks of the Techa River to watch a group of Westerners take radiation readings. Liquid high-level radioactive wastes from the chemical separation of plutonium were dumped directly into the river during the 1950s. Radiation levels today are 30 to 60 times higher than natural background. *Southern Urals, Chelyabinsk Region, Central Russia. May 23, 1992.*

Worldwide Cooperation

With the end of the Cold War, scientists and policymakers around the world are exchanging information and experiences in waste management, environmental cleanup, and the development of necessary technologies. For example, U.S. and Russian scientists are working together to develop chemical separation techniques for treating radioactive waste, technologies and methods for enhancing the characterization of sites requiring cleanup, and advanced thermal technologies for treating mixed waste. American scientists are learning a great deal from French, British, German, Japanese, Belgian, and Russian waste-vitrification projects. Representatives from the Department of Energy have visited the Capenhurst facility in Great Britain to learn more about the British experience in decontaminating and dismantling its gaseous-diffusion plant.

Safe Management of Nuclear Materials

In addition to international cooperation on environmental issues, the United States is working with other nations to reduce the proliferation threat of nuclear weapons and their components. Joint projects with Russia are helping the Russian government ensure that crucial materials for nuclear weapons are accounted for and well guarded. In 1991, the United States began a program to assist the Russians in dismantling their nuclear weapons and in managing their stockpiles of plutonium and highly enriched uranium as safely and securely as possible. Known as the Nunn-Lugar program after the senators who sponsored it, the program has already authorized funds for designing more secure, state-of-the-art storage vaults for Russian nuclear materials.

With the end of the Cold War, scientists and policymakers around the world are exchanging information in waste management, environmental cleanup, and technology development.

An International Perspective

The Department of Energy has also played a key supporting role in efforts to secure weapons-grade nuclear materials worldwide. The Department's Reduced Enrichment Research and Test Reactor program is aimed at eliminating international commerce in highly enriched uranium fuel for reactors–a material that could be diverted and used to make nuclear weapons. The Department is evaluating the renewal of this "take-back" policy through an environmental impact statement. The program asks the participating nations to return the highly enriched uranium reactor fuel that the United States originally supplied to them. In exchange, the United States would assist the participating countries in converting their reactors to use low-enrichment fuel, which is not suitable for weapons. The return of this nuclear material to the United States would significantly diminish the world's trade in weapons-usable uranium.

The dismantlement of surplus weapons and plants also can increase trust between the United States and its former Cold War adversaries. Surplus plutonium and highly enriched uranium in the United States and Russia are being opened up to international monitoring. Representatives of the International Atomic Energy Agency and the Russian government have already toured key U.S. production facilities at Hanford, Rocky Flats, and Oak Ridge.

New Attitudes

Nuclear weapons production has been veiled in secrecy worldwide since its beginnings half a century ago. During the Cold War, our national security strategy was based on deterring a large-scale Soviet attack by maintaining a large nuclear arsenal.

The definition of "national security" is now expanding to include other concerns: the environment, human health, the global economy, and the spread of nuclear weapons. Understanding these new missions, countries are sharing information and opening their once-secret facilities. Through such sharing, scientists, engineers, and policymakers can build trust; reduce environmental, safety, and health risks more effectively; and decrease the threat of nuclear weapons proliferation. Such alliances are forming the basis of a new world security.

The Soviet Nuclear Waste Legacy

The examples below provide perspective on the environmental legacy of nuclear weapons production in the former Soviet Union.

Techa River Contamination. The Techa River flows past the Mayak plutonium-production complex in the southern Ural Mountains. From 1949 to 1951, the Soviets pumped liquid high-level radioactive waste directly into the river. Without telling the residents why, the Soviet authorities evacuated about 8,000 people from 20 villages.

Lake Karachai. The contamination of the Techa River ended the practice of dumping high-level waste directly into the river. From 1948 until the late 1950s, engineers at Mayak dumped high-level waste into a small lake called Lake Karachai instead. Some 120 million curies of high-level waste (equal to about one-eighth of all the high-level waste generated by the U.S. complex) remains in Lake Karachai today. Workers filling in some of the reservoirs at Lake Karachai with concrete and dirt must operate their machines from shielded cabs. To this day, a person standing at some points on the lake's shore would receive a fatal dose of radiation in a few hours.

At times of drought, severely contaminated sediment from the lake's bottom dried out and was dispersed by the wind. The first such episode convinced the Soviets that this practice was unwise, and they began storing their wastes in aboveground tanks.

Mayak Waste-Tank Explosion. In 1957, an 80,000-gallon tank of high-level waste at Mayak exploded with a force of 5 to 10 tons of dynamite, heavily contaminating about 9,000 square miles. The average radiation dose received by some 10,000 people evacuated from the region was about 50 rem, 10 times the current annual limit for American radiation workers. Some 75 square miles remain uninhabitable today.

Waste Pumped Underground. As a result of the problems experienced at Mayak, other Soviet weapons-production sites began to pump high-level waste deep underground into rock formations that they believed would keep the waste from spreading or reaching the human environment. The quantity of waste thus disposed of was very large (about 1.5 billion curies), and most of the pumping occurred at the Siberian plutonium-production sites—Tomsk-7 on the Tom River and Krasnoyarsk-26 on the Yeni-sey River. The Soviets dumped other radioactive liquids into rivers and reservoirs near these sites.

The Arctic Ocean. Today, the Tom and Yenisey Rivers in Siberia are contaminated for hundreds of miles downstream. Some of the radioactive waste that was released into these rivers has ended up in the Arctic Ocean, where it has entered the ecosystem and endangered fisheries. Furthermore, the Soviet navy frequently dumped old submarine reactors and other highly radioactive materials directly into the Kara sea. Fallout from nuclear weapons testing on the arctic island of Novaya Zemlya has also contributed to the contamination of the Siberian arctic.

Closing the Circle on the Splitting of the Atom

VI. Transition to New Missions

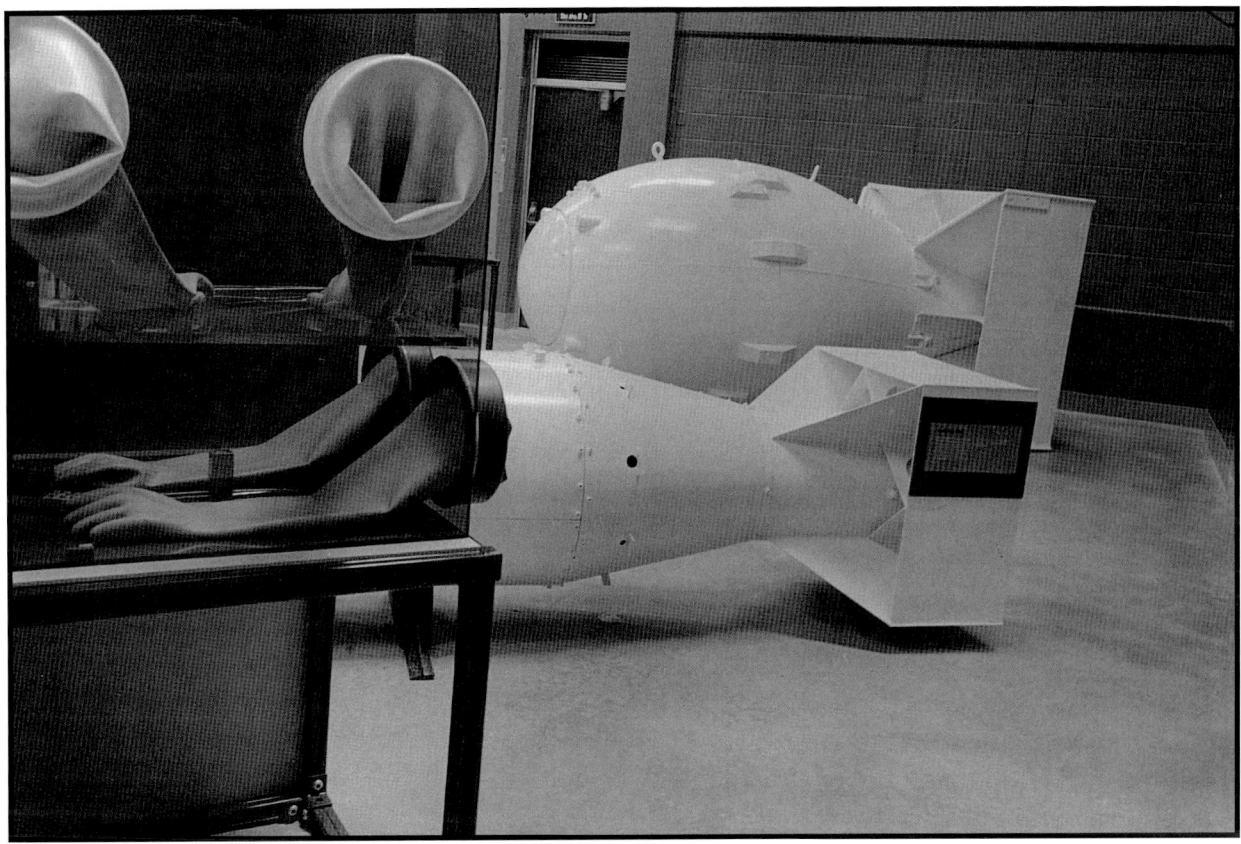

The first nuclear weapons. In the background is a duplicate of the casing for the Nagasaki plutonium bomb. In the middle distance is a duplicate of the casing for the Hiroshima uranium bomb. In the foreground, a model of a glovebox used for the handling of plutonium. *Bradbury Science Museum, Los Alamos, New Mexico. July 13, 1982.*

All of the major facilities in the U.S. nuclear weapons production complex were shut down in the late 1980s. For several different reasons, the end of production was quite sudden and largely unexpected. Incidents of mismanagement and contamination at U.S. nuclear weapons sites led to a series of Federal investigations into safety and environmental practices. These investigations pointed out that most of the Energy Department's weapons plants, built several decades ago, were at or near the end of their design life and unable to comply with current environmental and safety standards and regulations. Many operations were therefore discontinued while alternatives for weapons production were being considered. At about the same time, the Cold War began winding down, and in 1991 the Soviet Union collapsed, bringing the nuclear arms race of the Cold War to a sudden end.

The Department's increased emphasis on environmental management requires more than doing different tasks. The transition involves both engineering and institutional challenges.

Engineering Challenges

Because the shutdown of many facilities was unexpected, hazardous materials often remain there, sometimes having been left in the middle of a process step. Some scenes are reminiscent of the end of World War II, when workers in factories put their tools and papers down and never returned. In the case of the nuclear weapons complex, however, the work stopped before it was certain that it would not be restarted, and the tools and materials are extraordinarily hazardous. The Department currently maintains more than 20,000 buildings and structures, such as cooling towers for old nuclear reactors, that will eventually require decommissioning. Hence, the sheer number of facilities requires a systematic transi-

Closing the Circle on the Splitting of the Atom

Guard tower and security barriers at the Pantex Plant near Amarillo, in the Texas panhandle. Thousands of plutonium triggers from dismantled warheads are stored in bunkers at the site. *Pantex Plant, Texas. November 18, 1993.*

tion process for stabilization and preparation for decontamination and decommissioning.

Stabilizing and safeguarding these nuclear materials sometimes requires operating the facilities to prepare for cleanup. For example, workers are draining liquids from tanks to prevent leaks and processing chemicals to prevent fires and explosions. Because many of these tanks and chemicals contain plutonium and other nuclear materials, they require a level of safety and security at least as stringent as that for weapons production. In some cases renovations are necessary to bring the facilities into compliance with environmental and safety requirements.

The facilities in transition must first be stabilized. Until that step has been been completed, the buildings are not safe for cleanup. Not only is the stabilization work necessary for reasons of safety and worker protection, but it can also dramatically reduce the costs of long-term maintenance. Every dollar saved in annual maintenance costs is worth several dollars because the backlog of old facilities is so large that surplus facilities may wait years, and often decades, before they can be decontaminated.

Institutional Challenges

The institutional challenges of environmental management may be even more complex than the engineering tasks. To fulfill its new missions successfully, the Department must itself undergo a major institutional transformation. It must institute fundamentally different operating practices from those historically used to produce nuclear weapons. Complicated tasks in waste management and environmental restoration require better communication and coordination among facilities and operational divisions. Sustainable environmental and public-health policy depends on the involvement of citizens, State and local governments, Native American

All of the major nuclear weapons factories shut down in the late 1980s. At the time, the Department had made no plans to keep them shut down.

Tribes, and other Federal agencies. Such participation can be meaningful only with significant openness. Finally, cost-effective environmental management will require contracting reforms that reward efficiency and outstanding performance.

Need To Know

In the interest of national security, nuclear weapons workers generally knew only their particular jobs. As the Atomic Energy Commission said of the Manhattan Project:

> *Just as a man-of-war was compartmentalized to prevent a single torpedo from sending the vessel to the bottom, the [Manhattan Project] had been subdivided to prevent some indiscreet or disloyal individual from revealing the whole enterprise to the enemy.*

The Atomic Energy Commission used these words to describe the systemwide compartmentalization of knowledge deemed essential to building the first atomic bombs. The intentional narrowing of the field of knowledge, commonly called the "need-to-know" principle, asserts that there is no real need for individuals to have information beyond the minimum needed for their jobs. This approach to security pervaded the complex during the Cold War.

Knowledge of the whole picture is crucial to environmental cleanup. A narrow focus can hinder progress. It is now common practice in most industries to identify wastes that come from each part of a process and to determine how best to minimize or prevent their generation. If it did not understand these connections, the Department of Energy could create other problems while attempting to resolve the original concerns. For example, how should the Department manage new wastes that will be created from cleaning up contaminated soil, water, and buildings?

From Secrecy to Openness

Secrecy remains essential to maintaining the nuclear weapons stockpile. During the Cold War, a large amount of information about the nuclear weapons complex, including information on issues related to the environment, safety, and health, was withheld from the general public because of concerns about national security.

In keeping with the Clinton Administration's focus on government accountability, Energy Secretary Hazel O'Leary has begun an "openness initiative" to encourage informed and constructive citizen involvement.

This initiative has identified many types of information that no longer need to be kept secret to protect national security or prevent nuclear proliferation. Since December 1993, the Department has opened its files on previously unannounced nuclear tests, its data on inventories of plutonium and other material, and various information useful for more effective environmental management.

Two Statements by John Glenn, U. S. Senator, State of Ohio

1985: Hearing of the Governmental Affairs Committee, U.S. Senate, in Cincinnati, Ohio:

"Although most of us have become aware of the problems at Fernald only recently, the situation has existed for three long decades. And although we may not be able to do anything about the past releases of radiation from the plant, I strongly believe that the public has a right to know about such releases.

"We must see to it that what happened in the past is never repeated.... I'm fully aware of the economic and national security benefits the plant provides, but, as I said when I toured Fernald last month, while plants like Fernald are essential to the security of our country, we must see to it that the cost of that security does not include the health of our people."

1994: Confirmation hearing before the Governmental Affairs Committee, U.S. Senate, for Alice Rivlin as Director of the Office of Management and Budget:

"In 1985, the people at Fernald in Ohio wanted me to come out. They had problems there. I went out, not knowing how valid their concerns were, and found that they were very valid. We did General Accounting Office (GAO) studies then of the other spots in the nuclear weapons complex all over the country, some 11 States and 17 different major sites. Cleanup had been put away at that time. 'The Russians are coming; we have got to produce.' 'What are you going to do with the waste?' 'Put it out behind the plant.'

"... When we started this, the General Accounting Office estimated that to clean up the whole weapons complex was somewhere between $8 to $12 billion. Now the latest GAO estimate is $300 billion, if we can figure out how to do some of it, and over a 20- to 30-year period.... I am concerned about how we take care of these long-term items that are going to require a year-by-year effort.... Cleanup is not going to get cheaper as we go along and it is something that does have to be done because of the danger to our communities."

> ## David H. Nochumson: One Whistleblower's Story
>
> David H. Nochumson, manager of the Radioactive Air Emission Monitoring program at the Los Alamos National Laboratory in New Mexico, took his job very seriously. He knew that checking for radioactivity in air emissions was vital to protecting the health of people in and around the lab, and he knew the people whose health would be affected by how well he did his job.
>
> Nochumson, who had earned a Bachelor of Science degree in chemical engineering from Rutgers University and a Ph.D. in environmental engineering from Harvard, was hired by Los Alamos in 1978. In 1990, soon after he started a new assignment at the lab, he discovered that the lab had not complied with the requirements of the Clean Air Act for monitoring stack emissions of radioactive materials. He submitted a plan to bring the lab into compliance with the law, and a request for additional funding, but his supervisors did not implement it. He repeatedly explained the safety and noncompliance problems and the need for greater funding.
>
> One supervisor stated that he "could do away with" Nochumson's position; others told him to write only positive things about the lab and to seek counseling. Nochumson filed a complaint with the Department of Labor in June 1991 and stopped working in the same position.
>
> On September 27, 1994, a judge in the U.S. Department of Labor issued a decision in favor of Nochumson. The judge ordered the lab to reinstate Nochumson and pay him back wages and damages for lost work and emotional distress. The parties are now working together to settle this dispute.

The backlog of secret documents is monumental, roughly equivalent to a column of paper 3 miles high. Through its new Office of Declassification, the Department is working to open the records on such issues as highly enriched uranium, nuclear arsenals, health and safety, experiments with human beings and hundreds of other subjects. The Department is also reviewing the original secrecy rules mandated by the 1946 Atomic Energy Act.

Whistleblowers

An important part of the new policy of openness is encouraging "whistleblowers" to report lack of compliance with regulations, mismanagement, inefficiencies, fraud, and other problems. To highlight this initiative, in November 1993, Secretary of Energy Hazel R. O'Leary met with whistleblowers at a conference called "Protecting Integrity and Ethics." She has issued a call to "celebrate whistleblowers" and has promised to implement a policy of "zero tolerance" for reprisals against them.

Citizen Involvement

Many of the program's environmental questions cannot be answered with engineering solutions alone. Decisions about the most important questions can only be made through a national debate and cooperation among government officials; workers; contractors; all interested Federal, State, and Tribal parties; and informed citizens.

The important questions include the following:

- Who should decide the extent and schedule for cleaning up sites?

- How can public-health risks that might be incurred over hundreds or thousands of years be balanced against immediate risks to cleanup workers?

- What levels of risk are "acceptable" when they might affect large populations or extend over long periods?

- Who should oversee cleanup efforts and evaluate their results?

Contract Reform

The Department of Energy's current contracting system fulfilled the nation's Cold War priorities of designing, building, and testing nuclear weapons secretly and quickly. When production was the primary mission, one large contractor was responsible for virtually all services at each plant site, and that contractor was protected from most financial risks by the terms of the contract.

While appropriate for Cold War production, these types of contracts are not the best way to reach the new objectives of the Department. Contractors involved in environmental management activities will be required to demonstrate sound business practices and assume greater financial responsibility for activities within their control.

The Department of Energy, with fewer than 20,000 Federal workers and more than 140,000 contractor employees, has undertaken several initiatives to reform the way it

Secretary of Energy Hazel R. O'Leary with "whistleblowers" at a symposium entitled "Openness and Secrecy: Establishing Accountability in the Nuclear Age." The Secretary encourages dialogue with those who question the Department's operations and environmental compliance. She was a keynote speaker at this symposium. In the back row from left to right: Jim Vissar, Jeff Peters, Stephen Buckley, Government Accountability Project attorney Tom Carpenter, Casey Ruud, William S. Armijo, and John Brodeur. In the front row from left to right: Marlene Flor, Gaidine Oglesbee, Sonja Anderson, Secretary of Energy Hazel R. O'Leary, Gary Lekvold, Tim Powell, and Inez Austin. Kneeling: Jim Smith and Ed Bricker. *Washington, D.C. May 18, 1994.*

does business. Contract reform initiatives emphasize competition and the development of clear, objective performance criteria and measures. Performance-based incentives are focused on the accomplishment of the Department's strategic mission and reward contractors for fulfilling clear programmatic objectives. The Department is hiring more federal workers for onsite management and for verifying the performance of contractors. The Department has also begun to reallocate the financial and legal risks inherent in operating its sites in order to hold contractors more accountable.

As more facilities make the transition from production to environmental management, the Energy Department will continuously review contracting practices, competition, incentives, and penalties in order to support the paramount objectives of (1) protecting public health and the environment, (2) minimizing risks to workers, and (3) using public funds and resources efficiently and responsibility.

> The Department of Energy is removing the cloak of Cold War secrecy that has shrouded its nuclear weapons program for 50 years....
>
> The Cold War is over, and we're coming clean....
>
> In the old days, we decided, announced and then defended policy. In the new days we must engage the public, debate, decide, announce and then go forward.
>
> **From Secretary of Energy Hazel R. O'Leary's December 7, 1993, press conference announcing the openness initiative**

Closing the Circle on the Splitting of the Atom

VII. Looking to the Future

Planning for cleanup at the T Plant reprocessing "canyon" at Hanford. Engineers work on methods for decontaminating and eventually dismantling the world's oldest plutonium-separation plant. In the meantime, the facilities at the T Plant are being used to decontaminate equipment with high-activity contamination. *Hanford Site, Washington. July 11, 1994.*

Hundreds of thousands of people–machinists, physicists, engineers, cooks, truck drivers, secretaries, policymakers–worked in the nuclear weapons complex. From the beginnings of the Manhattan Project to the end of the Cold War, they accomplished extraordinary and unprecedented feats. As a nation, we are indebted to their ingenuity, enterprise, and plain hard work.

The nation is also indebted to those who brought to light the environmental, health, and safety problems throughout the complex–from "whistleblowers" at facilities to citizens living in the shadow of nuclear weapons sites to oversight committees in the Congress. The democratic values and rights championed by these individuals demonstrate exactly what the nation fought for during the Cold War. These rights and values continue to be vital to solving many of the problems and issues highlighted in this book. Only by building on such values can the Department of Energy make the kind of progress needed to get the job done properly.

Providing for Broad-Based Debate and Participation

The Department has demonstrated the value of linking its technical capabilities with democratic values. As major environmental projects began, public participation contributed much to the Department's efforts in stabilizing uranium-mill tailings and cleaning up numerous sites used during the Manhattan Project. The participation of state and local governments, regulatory agencies, Native American Tribes, and others also has been instrumental in writing environmental compliance agreements to provide meaningful and practical roadmaps for cleanup.

The Energy Department is committed to getting results with the tools available and to finding new technical solutions through research and development. It has dedicated significant resources to solving such problems as stabilizing high-level radioactive waste and removing or isolating soil and ground-water contamination.

Boxes containing low-level radioactive waste lie in a shallow land burial trench at the Savannah River Site. New methods for disposal of low-level waste are being developed by the Department. *Savannah River Site, South Carolina. January 7, 1994.*

In many cases, the most vexing problems cannot be addressed solely by science but will require a broad-based and informed public debate.

Strategy Before Action

The Department of Energy is evaluating how it will clean up its defense and nondefense facilities. This effort is based on its own recent experience and on lessons from the Environmental Protection Agency's Superfund program for hazardous waste. Both of these programs have made evident certain painful realities:

- Today's remediation technologies are often inadequate for fully solving many contamination problems, while innovative methods often encounter unexpected problems.

- Insufficient information is available for fully characterizing human and environmental risks.

- Few broadly accepted standards exist for determining "how clean is clean."

- The requirements for cleanup work often exceed available resources.

In light of these realities, the Department is trying to stabilize sites quickly, with a minimum of paper study, while investing in the development of more effective technologies. A better understanding of risks through the work of the new Environmental Management Office of Integrated Risk Management will help provide information to guide the program.

What are we doing today that will prompt another generation to say, "How could the scientists, policymakers, and environmental specialists not have seen the consequences of their actions?"

Looking to the Future

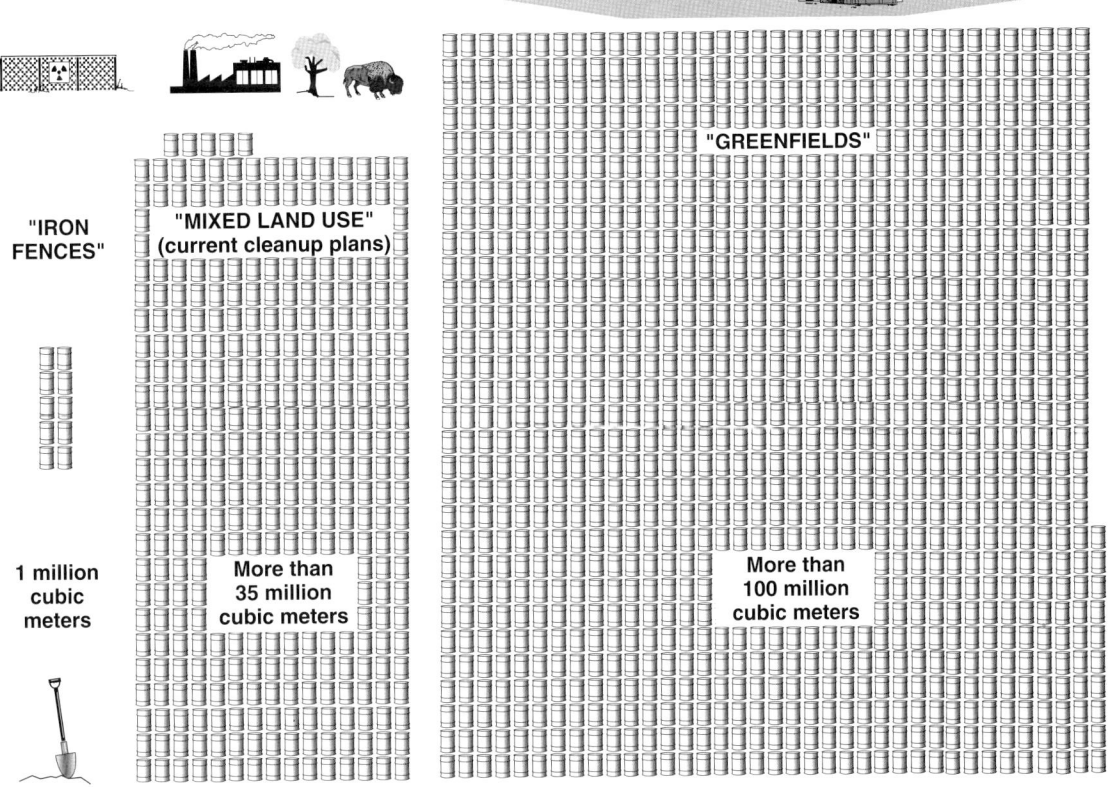

- Each drum represents 100,000 cubic meters of waste
- Based on current estimates from data collected for the 1995 Baseline Environmental Management Report (BEMR).

Dealing with the environmental legacy of the Cold War poses both technical and policy challenges. One imminent challenge is preparing the backlog of stored waste for disposal. The volume of waste currently in storage is nearly a third of the total amount of waste the Department has previously buried, virtually all of which was untreated before disposal. Most of this stored waste will require treatment prior to disposal. Consequently, the Department is undertaking an unprecedented campaign to design, site, construct, and operate facilities to comply with environmental laws. This endeavor will require not only developing new treatment technologies, but also working with regulators and other stakeholders to decide where and which treatment and disposal technologies should be used.

In the near future, difficult decisions will need to be made about what course of action the environmental restoration program should take. One of the principal issues will be: What will the land be used for after environmental restoration? Will it be sufficiently cleaned up for any purpose, including for a farm family that eats the crops grown on the land? If so, a large amount of contaminated soil may need to be dug up and many buildings will need to be dismantled, producing a large amount of waste requiring treatment or disposal. Another option is to remove less soil and limit dismantlement, producing substantially less waste. If this scenario is chosen, the Department will have to impose restrictions on how the land will be used after cleanup. For example, only industrial, nonresidential uses may be allowed for the land, and from the outset, environmental restoration would be directed at future industrial development, rather than "green-field" uses. Unfortunately, the United States has little experience with long-term land-use restrictions, which have a tradition of local, rather then federal, control. Whatever option is chosen, it is clear that the amount of waste produced by environmental restoration will depend largely on assumptions about how the land can be safely used in the future.

The Integrated Demonstration Site at the Savannah River Site contains 150 monitoring wells, some quite shallow, some as deep as 200 feet. These wells are used to chart the migration of contaminants through the water table and through different levels of soil and rock. Included within this site is the world's first horizontal injection well used for environmental remediation. *A-M Area, Savannah River Site, South Carolina. January 6, 1994.*

Brian Looney – Environmental Scientist

When Brian Looney was in high school during the early 1970s, he wanted to fix things and solve problems. The first Earth Day in 1970 gave him a sense of the connections among organisms and ecosystems, and of the importance of "thinking globally while acting locally."

In 1983, after receiving his Ph.D. in environmental engineering from the University of Minnesota, Looney came to the Department of Energy's Savannah River Site in South Carolina, where he is a research environmental engineer. Six years later he was put in charge of a major research effort to apply new technologies to ground-water cleanup. Looney focused on removing from the soil and ground water "plumes" of toxic chemicals that have spread from their sources.

His team completed the research project in September 1994. As Looney puts it, "We didn't ask for more funding. My bias is toward finding efficient methods and getting there with cost-effective research."

A major success of Brian's team has been the first application of horizontal drilling to environmental cleanup. These methods had been developed by private industry to enhance oil extraction and to install pipelines and cables. Although horizontal drilling costs more per foot drilled, it allows a much larger proportion of each borehole to be in close contact with the contaminated zone. The seven test wells at the Savannah River Site have removed about 2,500 gallons of toxic contaminants from about 14 million cubic feet of contaminated soil. Now some of the techniques developed by Looney and his team have been adapted to industrial cleanup. About 100 horizontal wells have since been drilled for environmental projects nationwide.

Other innovations flowing from Looney's research could dramatically improve the effectiveness of environmental cleanup. For example, his team developed efficient methods for introducing nutrients like phosphorus and nitrogen into soils to encourage bacteria that can break down pollutants. They have also demonstrated that in many cases the standard laboratory tests for pollution provide unnecessary precision at excessive cost. Larger numbers of less costly "field screening" samples can often lead to a better understanding of the extent of contamination.

These examples illustrate some elements of Looney's environmental philosophy. First, he says, "Let's work with Mother Nature instead of against her whenever possible." This means considering all the side effects of every cleanup project. Looney also keeps his eye on the "life-cycle" costs of his work and strives to keep them as low as possible. This philosophy and a new breed of dedicated environmental professionals like Looney are key to solving environmental problems that are as exciting and challenging as the Manhattan Project.

Looking to the Future

Addressing the worst environmental concerns first is necessary but not sufficient. The Department must also begin to reduce the backlog of environmental, safety, and health problems inherited from the Cold War. The experience gained so far suggests the following strategy:

- Where appropriate, stabilize radioactive materials to avoid accidents, the spread of contamination, and immediate risks to the public.
- Develop a thorough understanding of complicated waste and contamination problems instead of rushing into solutions that might have unexpected side effects.
- Develop effective technologies for cost-effective environmental work. An investment in technology can pay off with methods that could apply to other national and global waste problems.
- Where feasible and appropriate, ensure that site cleanup is part of a long-term solution rather than a hasty fix. Long-term solutions must take into account worker safety, public health risks, ecological values, and cost.
- Prepare for future uses of large portions of the more than 3,000 square miles reserved for the U.S. nuclear weapons complex. Much of that land is not significantly contaminated and can be returned to some level of public, industrial, or commercial use. Other sites can be released to the public after remediation or with appropriate limits on their uses.

Much of the 3,000 square miles reserved for the U.S. nuclear weapons complex is not significantly contaminated and can be returned to some level of public, industrial, or commercial use. Other sites can be released to the public after remediation or with appropriate limits on their uses.

A wastewater-treatment facility under construction at Hanford to treat low-level condensate from an evaporation process that reduces the volumes of high-level waste. Each of these three tanks has a capacity of 670,000 gallons. Because of the use of this facility, the Department has ceased discharges of contaminated waste water to the soil for the first time in more than 40 years. *Hanford Site, Washington. December 19, 1993.*

President Dwight D. Eisenhower warned that "the problem in defense is how far you can go without destroying from within what you are trying to defend from without." Meant as a warning against creating an all-powerful military-industrial complex, Eisenhower's statement is equally applicable to the environmental legacy of the Cold War.

Reconciling Democratic Involvement with Institutional Efficiency

Over the years, the Department's culture of secrecy and its history of contamination problems at nuclear weapons sites have profoundly affected both public attitudes and public opinion. Its credibility was among the lowest of any public institution. Ironically, many citizens who were previously shut out are now deluged with information and invitations to public meetings.

It will take more than meetings and paper to undo decades of mistrust. Only when the government and its contractors have earned again the trust of the public, the regulators, Native American Tribes, State and local governments, and public-interest groups, can there be meaningful progress. Involving outsiders in the Department's decision-making processes is only the first step along the path to trust. Ultimately, it will be actions that will define the Department. Every building and every waste site that is cleaned up will be another step forward. Trust will have to be built one relationship at a time, and it will take years to grow.

The Long-Term Vision for Environmental Management

Looking to the future of the Department's responsibilities in environmental management involves continually asking questions about the nature of the challenges to be faced and the ways of meeting them. The challenges can be met only with determined commitment over time. It is therefore imperative to think of strategic long-term goals. These goals cannot be established until questions like the following have been answered:

- How much environmental remediation does the nation want to "buy"?

- What are the savings from preventing future harm?

- What will the sites look like when the cleanup is complete?

- How do we control future land use?

- What is our obligation to future generations and to other species?

When these questions have been answered, environmental management programs can proceed with clear parameters and long-term goals.

What Might Future Generations Question?

Once removed from the fears and passions of the Cold War, many may find it easy to judge the actions of a generation now receding into history. Cooler minds today can, without great difficulty, conclude that many environmental problems could have been avoided through better housekeeping and waste management. But hindsight is not a useful lens through which to view our predecessors. The priority placed on weapons production over environmental protection was dictated by the imperatives of the time, just as the priorities of today are dictated by contemporary imperatives.

A question that haunts many who are involved in the Department's environmental management program is: "What are we doing today that will prompt another generation to say, 'how could those people – scientists, policymakers, and environmental specialists – not have seen the consequences of their actions?'" The question may be directed at current decisions about waste-disposal practices, or cleanup standards, or worker protection, or openness. No one can yet know what these future questions will be, much less the correct answers. Nonetheless, part of the inheritance of the people working on this new enterprise is a desire to look to the future and anticipate those questions.

Looking to the Future

Industrial safety sign at the Plutonium Finishing Plant. Signs like these help remind workers and managers to exercise sound safety procedures and keep them aware of the potential hazards associated with much of the work they do. *Hanford Site, Washington. July 11, 1994.*

If the intellectual giants of the Manhattan Project could not forsee all of the implications of their actions, it is particularly daunting for those involved in this new undertaking to consider what they might be missing in taking on the equally challenging task of cleaning up after the Cold War.

Closing the Circle on the Splitting of the Atom

The Department is building on a proud but troubled legacy – world-class scientific talent and a variety of environmental, safety and national security challenges inherited from the Cold War. To truly solve the problems left by the Cold War, the nation as a whole must commit itself to a sustained effort that will last for decades. Moreover, all of the people involved must look at the long-term consequences of current decisions in a way that, until now, has only rarely been done. Only then will future generations recognize this exciting but uncertain time as the beginning of the closing of the circle on the splitting of the atom.

> *All of the people involved must look at the long-term consequences of current decisions in a way that, until now, has only rarely been done.*

Closing the Circle on the Splitting of the Atom

GLOSSARY

Alpha particle. A particle consisting of two protons and two *neutrons*, given off by the *decay* of many elements, including *uranium, plutonium,* and *radon*. Alpha particles cannot penetrate a sheet of paper. However, alpha emitting *isotopes* in the body can be very damaging.

Americium. A manmade *transuranic element*; the next element following *plutonium* on the periodic table.

Atmospheric testing. The aboveground explosion of a nuclear device in order to test it or its effects.

Atom. The basic component of all matter. The atom is the smallest part of an element that has all of the chemical properties of that element. Atoms consist of a *nucleus* of protons and *neutrons* surrounded by electrons.

Atomic Energy Commission (AEC). The AEC was created by the United States Congress in 1946 as the civilian agency responsible for the production of nuclear weapons. The AEC also researched and regulated atomic energy. Its weapons production and research activities were given to the *Energy Research and Development Administration* in 1975, while its regulatory responsibility was given to the new Nuclear Regulatory Commission.

B Plant. The second *chemical separation* "*canyon*" built at the *Hanford Site* in Washington State for the *Manhattan Project* during World War II, the B plant was built between 1942 and 1945 and was used for *plutonium* recovery until 1956. Since then, it has had other uses. The code name "B" is arbitrary.

B Reactor. The world's first full-scale *plutonium production reactor*, the B reactor is located at the *Hanford Site* in Washington State. Construction on this reactor for the *Manhattan Project* started in 1943 and was completed in 1944. B reactor operated from 1944 to 1946 then from 1948 to 1968. The code name "B" is arbitrary.

Beta particle. A particle emitted in the *radioactive decay* of many *radionuclides*. A beta particle is identical with an electron. It has a short range in air and a low ability to penetrate other materials.

Calcine. A process that uses heat to reduce liquid *high-level waste* into a dry, powdery form. Also the powdered waste that results from this process.

Calutron. A device that uses an electromagnetic process to *enrich uranium*. Calutrons at the *Y-12 Plant* in *Oak Ridge* were used to *enrich uranium* for the *Manhattan Project*.

Canyon. A vernacular term for a *chemical separations plant*, inspired by the plant's long, high, narrow structure. Not all chemical separations plants are canyons.

Cesium. An element chemically similar to calcium. *Isotope* cesium-137 is one of the most important *fission products*, with a *half-life* of about 30 years.

Chain reaction. A self-sustaining series of nuclear *fission* reactions, when *neutrons* liberated by fission cause more fission. Chain reactions are essential to the functioning of *nuclear reactors* and weapons.

Chemical separation. Also known as *reprocessing;* a process for extracting *uranium* and *plutonium* from dissolved *irradiated targets* and *spent nuclear fuel* and *irradiated targets*. The *fission products* that are left behind are *high level wastes*.

Cladding. The outer layer of metal over the *fissile* material of a nuclear *fuel element*. Cladding on the Department of Energy's *spent fuel* is usually aluminum or zirconium.

Comprehensive Environmental Response, Compensation, and Liability Act (CERCLA). A Federal law, enacted in 1980, that governs the cleanup of hazardous, toxic, and radioactive substances. The Act and its amendments created a trust fund, commonly known as *Superfund*, to finance the investigation and cleanup of abandoned and uncontrolled hazardous waste sites.

Criticality. A term describing the conditions necessary for a sustained nuclear *chain reaction*.

Curie. The amount of radioactivity in 1 gram of the *isotope* radium 226. One curie is 37 billion *radioactive decays* per second.

Decay (radioactive). Spontaneous disintegration of the *nucleus* of an unstable *atom*, resulting in the emission of particles and energy.

Decay product. The *isotope* that results from the *decay* of an unstable *atom*.

Decommissioning. Retirement of a nuclear facility, including *decontamination* and/or dismantlement.

Decontamination. Removal of unwanted radioactive or hazardous contamination by a chemical or mechanical process.

Defense Waste Processing Facility. A *high-level-waste vitrification* plant built at the *Savannah River Site*.

Department of Energy (DOE). The cabinet-level U.S. Government agency responsible for nuclear weapons production and energy research and the cleanup of hazardous and radioactive waste at its sites. It was created from the *Energy Research and Development Administration* and other Federal Government functions in 1977.

Depleted uranium. *Uranium* that, through the process of *enrichment*, has been stripped of most of the *uranium 235* it once contained, so that it has more *uranium 238* than *natural uranium*. It is used in some parts of nuclear weapons and as a raw material for *plutonium* production.

Deuterium. A naturally occurring *isotope* of *hydrogen*. Deuterium is lighter than *tritium*, but twice as heavy as ordinary hydrogen. Deuterium is most often found in the form of *heavy water*.

Dose. As used here, a specific amount of *ionizing radiation* or toxic substance absorbed by a living being.

Dry cask storage. The storage of *spent nuclear fuel* without keeping it immersed in water.

Energy Research and Development Administration (ERDA). The agency created in 1975 to take over the weapons production and research responsibilities of the *Atomic Energy Commission*. ERDA was transformed, along with other Federal Government functions, into the cabinet-level *Department of Energy* in 1977.

Enrichment. The process of separating the *isotopes* of *uranium* from each other. Other elements can also be enriched. In the United States this is done using the *gaseous diffusion* process.

Enriched uranium. *Uranium* that, as a result of the process of *enrichment*, has more *uranium 235* than natural uranium.

Environmental contamination. The release into the environment of *radioactive*, hazardous and toxic materials.

Environmental Management. An Office of the *Department of Energy* that was created in 1989 to oversee the Department's waste management and environmental cleanup efforts. Originally called the Office of Environmental Restoration and Waste Management, it was renamed in 1993. Often abbreviated EM.

Environmental Protection Agency. A Federal agency responsible for enforcing environmental laws, including the *Resource Conservation and Recovery Act*; the *Comprehensive Environmental Response, Compensation and Liability Act*; and the *Toxic Substances Control Act*. The Environmental Protection Agency was established in 1970.

Epidemiology. The branch of medicine that studies the sources, distribution, and determinants of diseases and injuries in human populations.

Glossary

Evaporation pond. A pond constructed to hold liquid *radioactive* wastes so that the water can evaporate away, leaving behind the dissolved and suspended radioactive material.

Fernald plant. The *uranium* foundry built in the early 1950s to supply uranium for nuclear weapons production. Located near Fernald, Ohio, 20 miles northwest of Cincinnati. Known as the Feed Materials Production Center during the Cold War and now officially referred to as the Fernald Environmental Management Project.

Final assembly. The task of assembling a nuclear weapon from its component parts and sub-assemblies. This is done at the *Pantex Plant*.

Fissile. Capable of being split by a low-energy *neutron*. The most common fissile *isotopes* are *uranium 235* and *plutonium 239*.

Fission. The splitting or breaking apart of the *nucleus* of a heavy *atom* like *uranium* or *plutonium*, usually caused by the absorption of a *neutron*. Large amounts of energy and one or more *neutrons* are released when an atom fissions.

Fission products. The large variety of smaller *atoms*, including *cesium* and *strontium*, left over by the splitting of *uranium* and *plutonium*. Most of these atoms are *radioactive*, and they *decay* into other *isotopes*. There are more than 200 isotopes of 35 elements in this category. Most of the fission products in the United States are found in *spent nuclear fuel* and *high-level waste*.

Formerly Utilized Sites Remedial Action Program. A program to clean up privately owned facilities that were contaminated as a result of past nuclear weapons research and production. Many of these facilities did work for the *Manhattan Project*. Commonly referred to by its acronym, FUSRAP.

Fuel (nuclear). *Natural* or *enriched uranium* that sustains the *fission chain reaction* in a *nuclear reactor*. Also used to refer to the entire *fuel element*, including structural materials such as *cladding*.

Fuel element. Nuclear reactor fuel including both the *fissile* and structural materials, such as *cladding*, typically in the shape of a long cylinder.

Fusion. The process whereby the *nuclei* of lighter elements, especially the *isotopes* of *hydrogen* (*deuterium* and *tritium*) combine to form the *nucleus* of a heavier element with the release of substantial amounts of energy.

Gamma radiation. High-energy electromagnetic *radiation* emitted in the *radioactive decay* of many *radionuclides*. Gamma rays are similar to X-rays. They are highly penetrating.

Gaseous diffusion. The process used to make *enriched uranium* in the United States.

Geologic repository. A place to dispose of *radioactive* waste deep beneath the earth's surface.

Glovebox. A sealed box used to handle some *radioactive* materials with gloves attached to the wall. Often filled with an *inert gas* and fitted with a filtered ventilation system.

Half-life. The time it takes for one-half of any given number of unstable *atoms* to *decay*. Each *isotope* has its own characteristic half life. They range from small fractions of a second to billions of years. A general "rule of thumb" in health physics is that the hazardous period for a given isotope is 10 half-lives.

Hanford Site. A 570-square-mile Federal government-owned reservation in the desert of southeast Washington State. Established in 1943 as part of the *Manhattan Project*, the Hanford Site's chief mission has been the production of *plutonium* for use in nuclear weapons. Hanford is home to nine *production reactors* and four *chemical separation plants*.

Health physics. The science of *radiation* protection, established during the Manhattan Project.

Heavy water. Water that contains *deuterium atoms* in place of *hydrogen* atoms. Heavy water is used in the *Savannah River Site production reactors*.

Highly enriched uranium. *Uranium* with more than 20 percent of the *uranium 235* isotope, used for making nuclear weapons and also as *fuel* for some isotope-production, research, and power reactors. *Weapons-grade uranium* is a subset of this group.

High-level waste. Material generated by chemical *reprocessing* of *spent fuel* and *irradiated targets*. High-level waste contains highly *radioactive*, short-lived *fission products*, hazardous chemicals, and toxic heavy metals. High-level waste is usually found in the form of a liquid, a solid *saltcake*, a sludge, or a dry powdery *calcine*.

Hydrogen. The lightest element. Two of the three *isotopes* of hydrogen have been used in nuclear weapons: *deuterium* and *tritium*.

Idaho National Engineering Laboratory. An 893-square-mile Federal government-owned reservation in the eastern Idaho desert. The Idaho National Engineering Laboratory is the site of many research and test reactors and of the Idaho Chemical Processing Plant, where *spent nuclear fuel* from the U.S. Navy and from *research reactors* was *reprocessed*.

Inert gas. A gas that does not react chemically with other substances. The inert gases are helium, neon, argon, xenon, and radon. Also occasionally used inaccurately to refer to nitrogen.

Ionizing radiation. *Radiation* that is capable of breaking apart *molecules* or *atoms*. The splitting or *decay* of unstable *atoms* typically emits ionizing radiation.

Irradiate. To expose to *ionizing radiation*, usually in a *nuclear reactor*. *Targets* are irradiated to produce *isotopes*.

Isotopes. Different forms of the same chemical element that differ only by the number of *neutrons* in their *nucleus*. Most elements have more than one naturally occurring isotope. Many more isotopes have been produced in reactors and scientific laboratories.

K Reactor. A *plutonium* and *tritium production reactor* at the *Savannah River Site*, started in 1954 and shut down in 1988. The code name "K" is arbitrary.

K-25 Gaseous Diffusion Plant. The first full scale *gaseous diffusion* plant in the world, built in Oak Ridge, Tennessee, for the *Manhattan Project*. "K-25" is an arbitrary code name.

Lithium. The lightest metal, and the third lightest element. Lithium has two naturally occurring *isotopes*, lithium 6 and lithium 7. Lithium 6 *targets* are *irradiated* to manufacture *tritium*.

Los Alamos National Laboratory. The U. S. Government laboratory, established in 1943 as part of the *Manhattan Project*, that designed the first nuclear weapons. Located in northern New Mexico, about 60 miles north of Albuquerque.

Low-enriched uranium. *Uranium* that has been *enriched* until it consists of about 3 percent *uranium 235* and 97 percent *uranium 238*. Used as *nuclear reactor fuel*.

Low-level waste. A catchall term for any *radioactive* waste that is not *spent fuel, high-level,* or *transuranic waste*.

Manhattan Project. The U.S. Government project that produced the first nuclear weapons during World War II. Started in 1942, the Manhattan Project formally ended in 1946. The *Hanford Site*, the *Oak Ridge Reservation*, and the *Los Alamos National Laboratory* were created for this effort. Named for the Manhattan Engineering District of the U.S. Army Corps of Engineers.

Mined geologic disposal. See *geologic repository*.

Mixed waste. Waste that contains both chemically hazardous and *radioactive* materials.

Molecules. Larger structures formed by the bonding of *atoms*.

Glossary

N Reactor. The last *production reactor* built at the *Hanford Site*. The N reactor operated from 1963 through 1987. The code name "N" is arbitrary.

National Environmental Policy Act. A Federal law, enacted in 1970, that requires the Federal government to consider the environmental impacts of, and alternatives to, major proposed actions in its decisionmaking processes. Commonly referred to by its acronym, NEPA.

Natural uranium. *Uranium* that has not been through the *enrichment* process. It is made of 99.3 percent *uranium 238* and 0.7 percent *uranium 235*.

Neutron. A massive, uncharged particle that comprises part of the *nucleus*. *Uranium* and *plutonium atoms fission* when they absorb neutrons.. The *chain reactions* that make *nuclear reactors* and weapons work thus depend on neutrons. Manmade elements can be manufactured by bombarding other elements with neutrons in *production reactors*.

Nevada Test Site. A 1,350-square-mile area of the southern Nevada desert that has been the site of most of the U.S. *underground* and *atmospheric tests* since it opened in 1951. The site is some 65 miles northwest of Las Vegas.

Nonproliferation. Efforts to prevent or slow the spread of nuclear weapons and the materials and technologies used to produce them.

Nuclear reactor. A device that sustains a controlled nuclear *fission chain reaction*.

Nuclear weapons complex. The chain of foundries, *uranium enrichment* plants, reactors, *chemical separation* plants, factories, laboratories, assembly plants, and test sites that produces nuclear weapons. There were 16 major facilities in the U.S. nuclear weapons complex, located in 12 states.

Nucleus. The clump of protons and *neutrons* at the center of an *atom* that determine its identity and chemical and nuclear properties.

Oak Ridge. A 58-square-mile reservation near Knoxville, Tennessee. Oak Ridge was established as part of the *Manhattan Project* in 1943 to produce *enriched uranium*. Today it is the location of *K-25* and *Y-12* plants and the Oak Ridge National Laboratory (which was initially referred to by the arbitrary code name, "X-10.").

Pad. A flat concrete or asphalt surface used for the temporary storage of wastes. Its purpose is to keep wastes from leaching into the soil.

Pantex Plant. The United States' *final assembly* plant for nuclear weapons, located in the Texas panhandle near Amarillo.

PCBs. A group of commercially produced organic chemicals used since the 1940s in industrial applications throughout the *nuclear weapons complex*. Most notably, PCBs are found in many of the gaskets and large electrical transformers and capacitors in the *gaseous diffusion plants*. PCBs have been proven to be toxic to both humans and laboratory animals. "PCB" is an abbreviation of the full name, "polychlorinated biphenyls."

Plutonium. A manmade *fissile* element. Pure plutonium is a silvery metal that is heavier than lead. Material rich in the plutonium 239 *isotope* is preferred for manufacturing nuclear weapons, although any plutonium can be used. Plutonium 239 has a *half-life* of 24,000 years.

Plutonium residues. Materials left over from the processing of *plutonium* that contain enough plutonium to make its recovery economically worthwhile.

Plutonium pit. A vernacular term that refers to the spherical core of a thermonuclear weapon. This pit is the "trigger" of the primary portion of the weapon that, when compressed, reaches a critical mass and begins a sustained nuclear fission chain reaction.

Production reactor. A *nuclear reactor* that is designed to produce manmade *isotopes*. *Tritium* and *plutonium* are made in production reactors. The United States has 14 such reactors: nine at the *Hanford Site* and five at the *Savannah River Site*. Some *research reactors* are also used to produce *isotopes*.

PUREX. An acronym for Plutonium-Uranium Extraction, the name of the chemical process usually used to *reprocess spent nuclear fuel* and *irradiated targets*. Also refers to the first plant at the *Hanford Site* built to use this process. The PUREX plant operated from 1956 to 1972 and from 1983 to 1988.

Radiation. Energy transferred through space or other media in the form of particles or waves. In this document, we refer to *ionizing radiation*, which is capable of breaking up *atoms* or *molecules*. The splitting, or *decay*, of unstable *atoms* emits ionizing radiation.

Radioactive. Of, caused by, or exhibiting *radioactivity*.

Radioactivity. The spontaneous emission of *radiation* from the *nucleus* of an *atom*. *Radionuclides* lose particles and energy through the process of *radioactive decay*.

Radionuclide. A *radioactive* species of an *atom*. For example, *tritium* and *strontium 90 are* radionuclides of elements hydrogen and strontium.

Radon. A radioactive *inert gas* that is formed by the decay of radium. Radium is, in turn, a link in the decay chain of *uranium 238*. Radon, which occurs naturally in many minerals, is the chief hazard of *uranium mill tailings*.

Reprocessing. Synonymous with *chemical separation*.

Research reactor. A class of *nuclear reactors* used to do research into nuclear physics, reactor materials and design, and nuclear medicine. Some research reactors also produce *isotopes* for industrial and medical use.

Resource Conservation and Recovery Act (RCRA). A Federal law enacted in 1976 to address the treatment, storage, and disposal of hazardous waste.

Rocky Flats Plant. *Plutonium* processing and manufacturing plant located 21 miles northwest of Denver, Colorado. Rocky Flats made the plutonium triggers of nuclear weapons. Started operations in 1951. Now called the Rocky Flats Environmental Technology Site.

Saltcake. A cake of dry crystals of nuclear waste found in *high-level-waste* tanks.

Saltstone. A concrete-like material made with *low-level radioactive waste*.

Savannah River Site. A *plutonium* and *tritium* production site, established in 1950, covering 300 square miles along the Savannah River in South Carolina, near Augusta, Georgia. Five *production reactors* and two *chemical separation plants* are located here.

Shielding. Material used to block or absorb *radiation*. Often placed between sources of radiation and people or the environment.

Spent nuclear fuel. *Fuel elements* and *targets* that have been *irradiated* in a nuclear reactor.

Strontium. An element. *Isotope* strontium 90 is one of the most common *fission products*. It has a *half-life* of about 30 years. Strontium is chemically similar to calcium.

Superfund. A term commonly used to refer to the *Comprehensive Environmental Response, Compensation and Liability Act*.

Target. Material placed in a *nuclear reactor* to be bombarded with *neutrons*. This is done to produce new, manmade *radioactive* materials. Most important, targets of *uranium 238* are used to make *plutonium*, and targets of *lithium* are used to make *tritium*.

Glossary

Thermonuclear weapon. A nuclear weapon that uses *fission* to start a *fusion* reaction. Commonly called hydrogen bomb or "H-bomb."

Thorium. An element. Thorium is a byproduct of the *decay* of *uranium*.

Toxic Substances Control Act. A Federal law, enacted in 1976 to protect human health and the environment from unreasonable risk caused by exposure to or the manufacturing, distribution, use, or disposal of substances containing toxic chemicals.

Transport cask. A container used to transport *spent nuclear fuel* and other *radioactive* materials. Its purpose is to shield people from radiation while it is transported.

Transuranic elements. All elements beyond *uranium* on the periodic table. All of the transuranic elements are manmade.

Transuranic waste. Waste contaminated with *uranium 233* or *transuranic elements* having *half-lives* of over 20 years in concentrations of more than 1 ten-millionth of a *curie* of per gram of waste.

Tritium. The heaviest *isotope* of the element *hydrogen*. Tritium is three times heavier than ordinary hydrogen. Tritium gas is used to boost the explosive power of most modern nuclear weapons, inspiring the term, "hydrogen bomb." It is produced in *production reactors* and has a *half-life* of just over 12 years.

Tritium Facility. A plant at the *Savannah River Site* where *tritium* is separated from *lithium targets* and placed in capsules that are part of nuclear weapons.

Underground testing. Testing of a nuclear device or its effects by exploding it underground.

Uranium. The basic material for nuclear technology. It is a slightly *radioactive* naturally occurring heavy metal that is more dense than lead. Uranium is 40 times more common than silver.

Uranium hexafluoride. A gaseous form of *uranium* used in the *gaseous-diffusion enrichment* process.

Uranium mill. A plant where *uranium* is separated from ore taken from mines.

Uranium-mill tailings. The sandlike materials left over from the separation of *uranium* from its ore. More than 99 percent of the ore becomes tailings.

Uranium Mill Tailings Remedial Action Program. A program to reduce the hazards posed to the public by *uranium mill tailings*. The program was created by a Federal law passed in 1978. The *Department of Energy*'s *Office of Environmental Management* is responsible for carrying its implementation. Often referred to by its acronym, "UMTRA."

Uranium 233. A manmade *fissile isotope* of *uranium*.

Uranium 235. The lighter of the two main *isotopes* of *uranium*. Uranium 235 makes up less than 1 percent of the uranium that is mined from the ground. It has a *half-life* of 714 million years. Uranium 235 is the only naturally occurring *fissile* element.

Uranium 238. The heavier of the two main *isotopes* of *uranium*. Uranium 238 makes up over 99 percent of uranium as it is mined from the ground. It has a *half-life* of 4.5 billion years. It is not easily split by *neutrons*.

Vitrification. A process that stabilizes nuclear waste by mixing it with molten glass. The glass is poured into metal canisters, where it hardens into logs. Plants for vitrifying *high-level-waste* have been built in the United States at *West Valley*, New York, and the *Savannah River Site*.

Waste Isolation Pilot Plant. A *geologic repository* intended to provide permanent disposal deep underground for *transuranic wastes*. Located 2,150 feet underground in a salt bed near Carlsbad, New Mexico.

Closing the Circle on the Splitting of the Atom

Weapons-grade uranium. *Uranium* made up of over 90 percent of the *fissile uranium 235 isotope*.

Weldon Spring. Named for a town near St. Louis, Missouri, the Weldon Spring plant first performed many of the same *uranium* processing operations as the *Fernald plant*. The Weldon Spring plant operated from 1957 to 1966.

West Valley Demonstration Project. A plant near Buffalo, New York, used to demonstrate the *reprocessing* of *spent nuclear fuel* from commercial nuclear power plants. West Valley operated from 1966 to 1972. A *vitrification* plant for *high-level waste* has been built at the site.

Yellowcake. A common *uranium* compound, named for its typical color. Uranium is sent from the *uranium mill* to the refinery in this form.

Yucca Mountain. A site on, and adjacent to, the *Nevada Test Site* that is being examined to determine whether it is suitable for use as a *geologic repository* for the Department's *high-level wastes* and *spent fuel* from commercial nuclear reactors.

Y-12. A plant in *Oak Ridge*, Tennessee, built for the *Manhattan Project* to *enrich uranium* using *calutrons*. Today, this plant produces and stores components made of *enriched* and *depleted*

This granite block marks the location of buried radioactive materials that include wastes from Enrico Fermi's uranium-graphite pile, built for the Manhattan Project in 1942 under the University of Chicago's Stagg Field, then relocated to this area. The Fermi pile demonstrated the world's first man-made self-sustaining nuclear chain reaction. The caption on the marker reads: "CAUTION - DO NOT DIG Buried in this area is radioactive material from nuclear research conducted here 1943-1949. Burial area is marked by six corner markers 100 ft. from this center point. There is no danger to visitors. U.S. Department of Energy 1978." *Plot M, Palos Park Forest Preserve, Cook County Forest Preserve District, 20 miles southwest of Chicago, Illinois. November 5, 1995.*

For Further Reading

For those interested in the Department of Energy's environmental management program.

U.S. DEPARTMENT OF ENERGY PUBLICATIONS

The following publications can be obtained by calling the Environmental Management Information Center at 1-800-7EM-DATA (1-800-736-3282). Many are available at the Department's Public Reading Rooms around the country.

General Interest
These publications include a broad range of topics relating to the environmental management program and the history of the Department of Energy. They are meant to appeal to the general public.

Fehner, Terrence K., and Jack M. Holl. *Department of Energy, 1977-1994: A Summary History.* DOE/HR-0098. November 1994.

Gosling, F. G. *The Manhattan Project: Making the Atomic Bomb.* Washington, D.C.: U.S. Government Printing Office, September 1994.

Gosling, F. G., and Terence R. Fehner. *Closing the Circle: The Department of Energy and Environmental Management, 1942-1994*, DOE History Division, draft, March 1994.

U.S. Department of Energy. *EM Progress: A Report from the U.S. Department of Energy's Office of Environmental Management.* Published quarterly.

U.S. Department of Energy. *Committed to Results: DOE's Environmental Management Program.* DOE/EM-0152P. Washington, D.C.: U.S. Government Printing Office, April 1994.

U.S. Department of Energy. *Environmental Management Fact Sheets.* Washington, D.C.: U.S. Government Printing Office, August 1994.

U.S. Department of Energy. *Questions and Answers About the U.S. Department of Energy's Environmental Restoration Activities.* DOE/EM-0208. Washington, D.C.: U.S. Government Printing Office, November 1994.

U.S. Department of Energy. *Environmental Management 1994.* DOE/EM-0119. Washington, D.C.: U.S. Government Printing Office, February 1994. Note: *Environmental Management 1995* is to be published in February 1995.

Specific Topics
Many of these publications by the Department of Energy and its *contractors* are more technical in nature and may not appeal to the general-interest reader.

Bates, Mary Ellen, and Kathryn G. Norseth. *Site History of Savannah River.* Department of Energy History Division, January 1993.

Belanger, Dian O. *Site History of the Oak Ridge Reservation.* Department of Energy History Division, January 1993.

Bradley, Donald J., and K. J. Schneider. *Radioactive Waste Management in the USSR: A Review of Unclassified Sources, 1963-1990.* PNL-7182. Richland, Washington: Pacific Northwest Laboratory, March 1990.

Dudgeon, Ruth A. *Site History of the Pantex Plant.* Department of Energy History Division, January 1993.

Gerber, Michele Stenhejem. *The Hanford Site: An Anthology of Early Histories.* WHC-MR-0435. Richland, Washington: Westinghouse Hanford Co., October 1993.

Gerber, Michele Stenhejem. *Legend and Legacy: Fifty Years of Defense Production at the Hanford Site.* WHC-MR-0293, Rev. 2. Richland, Washington: Westinghouse Hanford Co., September 1992.

Goldman, David I. *Site History of Idaho National Engineering Laboratory.* Department of Energy History Division, January 1993.

Golightly, Eric J. *Site History of the Fernald Environmental Management Project.* Department of Energy History Division, January 1993.

Keystone Center. *Interim Report of the Federal Facilities Environmental Restoration Dialogue Committee: Recommendations for Improving the Federal Facilities Environmental Restoration Decision-Making and Priority-Setting Processes.* Washington, D.C.: U.S. Government Printing Office, February 1994.

U.S. Department of Energy. *Earning Public Trust and Confidence: Requisites for Managing Radioactive Wastes, Final Report of the Secretary of Energy Advisory Board Task Force on Radioactive Waste Management.* Washington, D.C.; U.S. Government Printing Office, November, 1993.

U.S. Department of Energy. *Long Range Master Plan for Defense Transuranic Waste Program.* DOE/WIPP 88-028. Carlsbad, New Mexico, December 1988.

U.S. Department of Energy. TRU *Waste Acceptance Criteria for the Waste Isolation Pilot Plant,* WIPP/DOE-069, Rev. 4. Carlsbad, New Mexico, December 1991.

U.S. Department of Energy. *Interim Mixed Waste Inventory Report: Waste Streams, Treatment Capacities and Technologies.* DOE/NBM-1100. Washington, D.C.: U.S. Government Printing Office, April 1993.

U.S. Department of Energy. *Spent Fuel Working Group Report on Inventory and Storage of the Department's Spent Nuclear Fuel and Other Reactor Irradiated Nuclear Materials and their Environmental, Safety and Health Vulnerabilities.* Washington, D.C.: U.S. Government Printing Office, November 1993.

U.S. Department of Energy. *Making Contracting Work Better and Cost Less. Report of the Contract Reform Team.* DOE/S-0107. Washington, D.C.: U.S. Government Printing Office, February 1994.

U.S. Department of Energy. *Annual Report on Waste Generation and Waste Minimization Progress 1991-2.* DOE/S-0105. Washington, D.C.: U.S. Government Printing Office, February 1994.

U.S. Department of Energy. *Waste Minimization/Pollution Prevention Crosscut Plan for 1994.* DOE/S-0094P. Washington, D.C.: U.S. Government Printing Office, February 1994.

U.S. Department of Energy. *Plan of Action to Resolve Spent Nuclear Fuel Vulnerabilities, Phases I, II, and III.* Washington, D.C.: U.S. Government Printing Office, February, April, and October 1994.

U.S. Department of Energy. *Integrated Data Base for 1993: U.S. Spent Fuel and Radioactive Waste Inventories, Projections and Characteristics.* DOE/RW-0006, Rev. 9. Washington, D.C.: U.S. Government Printing Office, March 1994.

U.S. Department of Energy. *Department of Energy Programmatic Spent Nuclear Fuel Management and Idaho National Engineering Laboratory Environmental Restoration and Waste Management Programs Draft Environmental Impact Statement.* DOE/EIS-0203-D. Washington, D.C.: U.S. Government Printing Office, June 1994.

U.S. Department of Energy. *Plutonium Working Group Report on Environmental, Safety and Health Vulnerabilities Associated with the Department's Plutonium Storage.* DOE/EH-0415. Washington, D.C.: U.S. Government Printing Office, September 1994.

U.S. Department of Energy. *Chemical Safety Working Group Report.* (three volumes.) DOE/EH-0396P. Washington, D.C.: U.S. Government Printing Office, September 1994.

U.S. Department of Energy. *Management Response Plan for the Chemical Safety Working Group Report.* (two volumes.) DOE/EH-0396P. Washington, D.C.: U.S. Government Printing Office, September 1994.

U.S. Department of Energy. *Materials in Inventory Report.* November 1994.

U.S. Department of Energy. *Plutonium Vulnerability Management Plan.* DOE/EM-0199. Washington, D.C.: U.S. Government Printing Office, January 1995.

Zonzon, Joan M. *Site History of Los Alamos.* Department of Energy History Division, January 1993.

PUBLICATIONS FROM OTHER SOURCES

These resources, including *books, pamphlets, and newsletters,* are available from their publishers. Many of them can be found in municipal or university libraries as well. Some of these authors present viewpoints inconsistent with those of the Department of Energy, or include technical data that the Department has not endorsed.

General Interest
These publications cover broad areas of the Department of Energy's nuclear weapons and environmental management programs. Most are meant to appeal to the general public.

Beyond the Bomb: Dismantling Nuclear Weapons and Disposing of their Radioactive Wastes. Peter Gray, editor. San Francisco: The Tides Foundation; and Seattle: Nuclear Safety Campaign, January 1994.

Del Tredici, Robert. *At Work in the Fields of the Bomb.* New York: Harper and Row, 1989.

Facing Reality: The Future of the U.S. Weapons Complex. Peter Gray, editor. San Francisco: The Tides Foundation; and Seattle: Nuclear Safety Campaign, May 1992.

Gerber, Michele Stenhejem. *On the Home Front: The Cold War Legacy of the Hanford Nuclear Site.* University of Nebraska Press, 1992.

Groves, Leslie. *Now It Can Be Told: The Story of the Manhattan Project.* New York: Da Capo Press, 1962.

Hewlett, Richard G., and Oscar E. Anderson, Jr. *The New World, 1939-1946, Volume I of A History of the United States Atomic Energy Commission.* University Park: Pennsylvania State University Press, 1962.

Hewlett, Richard G., and Francis Duncan. Atomic Shield, 1947-1952, Volume II of *A History of the United States Atomic Energy Commission.* University Park: Pennsylvania State University Press, 1969.

Hewlett, Richard G., and Jack M. Holl. Atoms for Peace and War, 1953-1961, Volume III of *A History of the United States Atomic Energy Commission.* Berkeley: University of California Press, 1989.

Johnson, Charles W., and Charles O. Jackson. *City Behind a Fence: Oak Ridge,* Tennessee, 1942-1946. Knoxville: University of Tennessee Press, 1981.

League of Women Voters Education Fund. *The Nuclear Waste Primer, A Handbook for Citizens,* 1993 Revised Edition. Washington, D.C.: U.S. Government Printing Office, 1993.

Lenssen, Nicholas. *Nuclear Waste: The Problem That Won't Go Away.* Worldwatch Paper 106. Washington, D.C.: Worldwatch Institute, December 1992.

May, John. *The Greenpeace Book of the Nuclear Age: The Hidden History, The Human Cost.* Toronto: McClelland & Stewart, 1989.

McPhee, John. *The Curve of Binding Energy.* New York: Farrar, Strauss and Giroux, 1974.

Murray, Raymond L. *Understanding Radioactive Waste.* Columbus, Ohio: Battelle Memorial Press, 1989.

National Research Council, Commission on Science, Math and Resources. *The Nuclear Weapons Complex: Management for Health, Safety and the Environment.* Washington, D.C.: National Academy Press, 1989.

Robinson, Marilynne. *Mother Country.* New York: Farrar, Straus & Giroux, 1989.

Rhodes, Richard. *The Making of the Atomic Bomb.* New York: Simon and Schuster, 1986.

Shulman, Seth. *The Threat at Home.* Boston: Beacon Press, 1992.

Teller, Edward, with Allen Brown. *The Legacy of Hiroshima.* Garden City, New York: Doubleday and Co., 1962.

Udall, Stewart L. *The Myths of August: A Personal Exploration of our Tragic Cold War Affair with the Atom.* New York: Pantheon Books, 1994.

U.S. Congress, Congressional Budget Office. *Cleaning Up the Department of Energy's Nuclear Weapons Complex.* Washington, D.C.: U.S. Government Printing Office, May 1994.

U.S. Congress, Office of Technology Assessment. *Complex Cleanup: The Environmental Legacy of Nuclear Weapons Production.* OTA-O-484. Washington, D.C.: U.S. Government Printing Office, February 1991.

U.S. Congress, Office of Technology Assessment. *Hazards Ahead: Managing Cleanup Worker Health and Safety in the Nuclear Weapons Complex.* OTA-BP-O-85. Washington, D.C.: U.S. Government Printing Office, February 1993.

Specific Topics
These publications cover more specific topics. Many of them are more technical in nature and may not appeal to the general-interest reader.

Carter, Luther. *Nuclear Imperatives and Public Trust: Dealing with Radioactive Waste.* Baltimore: Resources for the Future, 1987.

Chow, Brian G. and Kenneth A. Solomon. *Limiting the Spread of Weapons-Usable Fissile Materials.* Santa Monica, California: National Defense Research Institute, RAND Corporation, November 1993.

Controlling the Atom in the 21st Century. David T. O'Very, Christopher E. Paine and Dan W. Reicher. Boulder, Colorado: Westview Press, 1993.

Finnamore, Barbra. "Regulating Hazardous and Mixed Waste at Derpartment of Energy Nuclear Weapons Facilities." *Harvard Environmental Law Review,* Vol. 9, No. 1, 1985.

Geiger, H. Jack, M.D., et al. *Dead Reckoning: A Critical Review of the Department of Energy's Epidemiologic Research.* Washington, D.C.: Physicians for Social Responsibility, 1992.

Hacker, Barton C. *The Dragon's Tail: Radiation Safety in the Manhattan Project, 1942-1946.* Berkeley: University of California Press, 1987.

Hidden Dangers. Anne H. Erhlich and John W. Birks, editors. San Francisco: Sierra Club Books, 1990.

International Physicians for the Prevention of Nuclear War and the Institute for Energy and Environmental Research. *Plutonium: Deadly Gold of the Nuclear Age.* Cambridge, Massachusetts: International Physicians for the Prevention of Nuclear War; and Takoma Park, Maryland: Institute for Energy and Environmental Research, 1992.

Lamont, Lansing. *Day of Trinity.* New York: Atheneum, 1965.

Lanouette, William. *Tritium and the Times. How the Nuclear Weapons-Production Scandal Became a National Story.* Research Paper R-1. Cambridge, Massachusetts: The Joan Shorenstein Barone

Center for Press Politics and Public Policy. Harvard University, John F. Kennedy School of Government. May 1990.

Makhijani, Arjun, and Scott Saleska. *High Level Dollars, Low-Level Sense: A Critique of Present Policy for the Management of Long-Lived Radioactive Wastes and Discussion of an Alternative Approach.* Takoma Park, Maryland: Institute for Energy and Environmental Research, 1992.

Makhijani, Arjun. *Fissile Materials in a Glass Darkly: Technical and Policy Aspects of the Disposition of Plutonium from Dismantled Nuclear Weapons.* Takoma Park, Maryland: Institute for Energy and Environmental Research, November 1994.

National Research Council, Commission on the Biological Effects of Ionizing Radiation. *Health Effects of Exposure to Low Levels of Ionizing Radiation (BEIR V Report.).* Washington, D.C.: National Academy Press, 1990.

National Research Council. *Management and Disposition of Excess Weapons Plutonium.* Washington, D.C.: National Academy Press, 1994.

National Research Council, Committee to Review Risk Management in the DOE's Environmental Remediation Program. *Building Consensus Through Risk Assessment and Management of the Department of Energy's Environmental Remediation Program.* Washington, D.C.: National Academy Press, 1994.

U.S. Congress, Office of Technology Assessment. *Long-Lived Legacy: Managing the High-Level and Transuranic Waste at the Department of Energy Nuclear Weapons Complex.* OTA-BP-O-83. Washington, D.C.: U.S. Government Printing Office, May 1991.

U.S. Congress, Office of Technology Assessment. *Dismantling the Bomb and Managing the Nuclear Materials.* OTA-O-572. Washington, D.C.: U.S. General Accounting Office, September 1993.

U.S. General Accounting Office. *Nuclear Waste: Hanford Single-Shell Tank Leaks Greater than Estimated.* GAO/RCED-91-177. Washington, D.C.: U.S. General Accounting Office, August 1991.

U.S. General Accounting Office. *Nuclear Waste: Defense Waste Processing Facility ñ Cost, Schedule and Technical Issues.* GAO/RCED-92-183. Washington, D.C.: U.S. General Accounting Office, June 1992.

U.S. General Accounting Office. *Nuclear Health and Safety: Examples of Post-World War II Radiation Releases at U.S. Nuclear Sites.* GAO/RCED-94-51FS. Washington, D.C.: U.S. General Accounting Office, November 1993.

U.S. General Accounting Office. *Nuclear Nonproliferation: Concerns with U.S. Delays in Accepting Foriegn Research Reactors' Spent Fuel.* GAO/RCED-94-119. Washington, D.C.: U.S. General Accounting Office, March 1994.

U.S. General Accounting Office. *Nuclear Cleanup: Completion of Standards and Effectiveness of Land Use Planning are Uncertain.* GAO/RCED-94-144. Washington, D.C.: U.S. General Accounting Office, August 1994.

U.S. General Accounting Office. *Nuclear Waste: Much Effort Needed to Meet Federal Facility Compliance Act Requirements.* GAO/RCED-94-179. Washington, D.C.: U.S. General Accounting Office, May 1994.

U.S. General Accounting Office. *Nuclear Waste: Foreign Countries' Approaches to High Level Waste Storage and Disposal.* GAO/RCED-94-172. Washington, D.C.: U.S. General Accounting Office, August 1994.

U.S. General Accounting Office. *Department of Energy: Management Change Needed to Expand Use of Innovative Cleanup Technologies.* GAO/RCED-94-205. Washington, D.C.: U.S. General Accounting Office, August 1994.

U.S. General Accounting Office. *Nuclear Waste: Comprehensive Review of the Disposal Program Is Needed.* GAO/RCED-94-299. Washington, D.C.: U.S. General Accounting Office, September 1994.

U.S. General Accounting Office. *Nuclear Health and Safety: Consensus on Acceptable Radiation Risk to the Public.* GAO/RCED-94-190. Washington, D.C.: U.S. General Accounting Office, September 1994.

U.S. General Accounting Office. *Nuclear Waste: Further Improvement Needed in the Hanford Tank Farm Maintenance Program.* GAO/RCED-95-29. Washington, D.C.: U.S. General Accounting Office, November 1994.

Werner, James D. "Secrecy and its Effect on Environmental Problems in the Military: An Engineer's Perspective." *New York University Environmental Law Journal.* Vol. 2, pages 351-9, 1993

Periodicals

Bulletin of the Atomic Scientists. Chicago: Education Foundation for Nuclear Science. Published monthly.

Defense Cleanup. Arlington, Virginia: Pasha Publications. Published weekly.

Energy Daily. Washington, D.C.: King Publishing Group. Published daily.

Inside Energy. New York: McGraw-Hill, Inc. Published weekly.

Nuclear Waste News. Silver Spring, Maryland: Business Publishers. Published weekly.

The Radioactive Exchange. Washington, D.C.: Exchange Publications. Published 23 times a year.

Weapons Complex Monitor: Waste Management and Cleanup. Washington, D.C.: Exchange Publications. Published biweekly.